Norfolk and Western Six-Eleven

3 Times A Lady

photo by Kenneth L. Miller

BY TIMOTHY R. HENSLEY AND KENNETH L. MILLER

ACKNOWLEDGEMENTS

This work, while original and mostly new for this edition, is largely excerpted from a revised edition of the 2000 book *Norfolk and Western Class J: The Finest Steam Passenger Locomotive*. Please step up and support continuing preservation of 611, 1218 and other representations displayed in Virginia Museum of Transportation's noteworthy railroad collection.

Our thanks to Norfolk Southern Corporation for use of the many fine company images that their former N&W photographers recorded during the steam era. Archivist Jennifer McDaid and current NS digital lens artist Casey Thomason were most helpful in that regard.

If someone's name appears in the text, rest most assuredly that they rendered assistance to the authors. Our thanks to retired teacher Pam Eller for proofreading. Thanks to Beth Kressler Miller who offered support and also was a proofreader. Carl Jensen, retired Manager-Steam Operations provided valuable material and insight.

Among those that worked behind the scenes were Jim Detty, who provided Scioto Division notes about the 611, and Larry Evans, who served as a liaison to the Norfolk & Western Historical Society. Colonel "Bud" Jefferies provided some additional information on the tandem rods. Charlie Schlotthober provided information on the brakeman's cupola and Ron Davis of N&WHS provided support. Harry Bundy helped clarify some information on the last 1959 excursion on the Norfolk Division. A special nod goes to eminent Norfolk and Western historian, the late Art Bixby who assembled and organized a vast amount of information.

And, especially, to the many folks who designed, built, maintained and operated the N&W fleet of "Modern Coal Burning Steam Locomotives." Fortunately, we interviewed about 100 "Folks Along The Line" who were involved with that golden era and share some of that here. Most have already passed on to their reward, but their voices live on through prized oral history tapes, some recorded as long as 45 years ago. Little could they imagine that their work experiences – some started 90 years ago – would matter to anyone now.

One of the more colorful characters that co-author Hensley encountered was J. L. "Lake" Waldron, a retired Norfolk Division engineer who lived up a cove near Troutville, Va. Not long after the introductions took place, he thundered at his wife out in the kitchen: "Darcy! Bring this boy two sausage & egg biscuits and a cup of coffee!" This was long before McDonald's came into vogue.

Protesting that he already had breakfast, Hensley was shocked when Waldron looked at him point blank and said, "If ye'ain't a'gonna eat sump'n, I ain't a'gonna tell ye'nothin'!" Naturally, the co-author left there well-nourished and with a wealth of home-spun knowledge. Such are the spoils of southern hospitality!

First Edition 2015
The contents of this edition are excerpted from a revised edition of
Norfolk and Western Class J: The Finest Steam Passenger Locomotive

Library of Congress Control Number: 2015943390
ISBN: 978-0-9899837-1-6
Layout and Design by Kenneth L. Miller, Miller Design & Photography, Salem, Va.

Published 2015
by Pocahontas Productions
1706 Chestnut Street
Kenova, WV 25530

Foreword by Preston Claytor

December 7, 1994 was a sad day. This was the day that Norfolk and Western 611 made its final trip from Salisbury, N. C. to Roanoke, Va., where it would again be interred in the, by then, Virginia Museum of Transportation. A number of us present at Shaffers Crossing in Roanoke that dreary, late fall day quietly vowed that we would not allow this to be the last time under steam for 611. Perseverance did, in fact, pay off.

Trackside for what was thought to be 611's last trip were two true friends of the locomotive, Timothy R. Hensley and Kenneth L. Miller. Miller took pictures while Hensley chose to watch and burn the image into his memory. These actions tell the tale of two great historians and a combination of talents including, writing skills, graphic arts and photography. Little did I know on that dismal day that I would turn to these two to author the history of 611 less than 20 years later.

Authors Miller (left) and Hensley alongside a tired and dusty 611 at Lynchburg, Va, December 7, 1994. *Jeffry L. Sanders photo*

One of my first official acts as chairman of FireUp611! was to ask this talented twosome to write that history.

The list of accomplishments of Tim Hensley and Ken Miller are significant. Tim authored the text for *Steam, Steel & Stars*, the first book published featuring the photography of famed photographer of N&W steam locomotives, O. Winston Link. Ken is the author of *Norfolk and Western Class J: The Finest Steam Passenger Locomotive*. Together, under their Pocahontas Productions banner, the two would collaborate on *Cass Scenic Railroad: Fifty Years a State Park – A Century of Steam on Bald Knob*.

What a team this pair has turned out to be. Tim Hensley, a journalism graduate of Marshall University has written professionally for the Brotherhood of Locomotive Engineers as well as for CSX Transportation's corporate newspaper. Entering service as a brakeman for Chessie System, Hensley was soon promoted to engineer and rose to the rank of Resident Vice-President for his home state of West Virginia. Now retired after spending the last 13 years at the throttle of Amtrak's *Cardinal*, he is the owner of *The Trainmasters House* Bed & Breakfast in Kenova, W. Va.

Tim continues to manage one of the finest collections of Norfolk and Western Railway memorabilia that can be found.

Ken Miller is a skilled photographer and graphic artist who is the author of the aforementioned book which is the definitive history of N&W's Class J locomotives. His pictures are numerous in this publication. Always the historian, Ken continues to gather and unearth new historical facts for a future updated edition. Ken is also an extremely talented graphic artist. One of his skills is the ability to reproduce the exact fonts and letters used to letter steam locomotives and tenders. Ken produced and lead the application of all of the stenciling utilized during the 2015 painting of N&W 611. A testament to his talents comes from D. Scott Lindsay, 611's Chief Mechanical Officer and owner of Steam Operations Corporation, who has contracted Ken for preparing stenciling on other locomotive restorations.

Born eight months after 611's final trips in 1959, I was not around to experience 611's first life. Fortunately I was able to play a role in the

The weather is about as gloomy as the watcher's mood on December 7, 1994. The 611 leads her dirty, ragged train through Montvale, Roanoke is next – and last – stop. Those who witnessed the events felt it would likely be the last time they would witness the 611 under steam. Many had tears in their eyes. *Kenneth L. Miller photo*

Fireman Preston Claytor is taking a break during the trip back to Roanoke in August 1982. *Both: Kenneth L. Miller*

1982 rebuild of the locomotive at the Southern Railway Steam Shop in Birmingham, Ala. Over the next 12 years, I had the great pleasure to ride behind, work on, fire and run the locomotive that the late, great James A. Bistline dubbed the "Mighty J".

In 2010, we all saw the coming of the Norfolk Southern's *21st Century Steam* and knew that 611 had to participate. Working behind the scenes a select group of us convinced 611's owners that restoring the locomotive could be a game changer for the Virginia Museum of Transportation, Inc.

Using 611 as a backdrop on a frigid February day in 2013, we announced that an exploratory committee would be formed to determine if the big Class J could be restored to operation. My father, a lawyer, always said: "Never ask a question that you don't already know the answer to..."

In June of 2013, we announced that we would undertake a capital campaign and with the generous donation of $1.5 million by Norfolk Southern in October of that year, we knew that we would achieve our goal and that 611 would be rebuilt!

My role in the project from inception through May 15, 2015 was Chairman of the FireUp611! committee, later project director and excursion director. We told our volunteers that we weren't just restoring history but that we were actually writing the next chapter. It was after I stated this at one of our training sessions that I realized that we needed someone to document the history of Norfolk and Western Ry. 611.

Enter Ken Miller and Tim Hensley. Of all the decisions that had to be made for the restoration project, the selection of these two historians to author the 611 story was the classic no-brainer.

In *3 Times a Lady* the reader will enjoy a mix of interesting personal stories as well as the history of 611's three careers. Hensley and Miller have done a great job of uncovering new evidence about 611. One of the most interesting is the documentation of my uncle, W. Graham Claytor, Jr. and father, Robert B. Claytor's roles in saving 611 from the scrappers following the October 1959 fan trips. The accounting by the authors exactly matches the family stories told around the dinner table for years.

I am personally honored to have played a small role in the third segment of 611's history. However we would not have had the opportunity without the efforts of my famous father and uncle. I treasure autographed and inscribed manuscripts above the standard book. In writing this foreword I turned to the collection for some personal notes. The first is from O. Winston Link to my father in *Steam Steel & Stars*:

> *To Bob Claytor, Merger of the N&W with Southern was a big deal and important but only you could have managed to get 611 & 1218 back where they belong and doing what they should be doing. Winston*

From the same manuscript, Tim Hensley wrote:

> *Preston, I hope this book will underscore an appreciation of your Norfolk & Western heritage and serve as a continuing reminder of our friendship. Tim Hensley January 11, 1988.*

And finally Ken Miller wrote to me in *Norfolk & Western Class J: The Finest Steam Passenger Locomotive*:

> *Preston - Keep the faith. Ken Miller*

Ken, we all did. And herewith is presented the most factual account of the locomotive that could be written.

J. Preston Claytor
Village of Golf, Fla.
October 2015

December 7, 1994. The 611 is only a mile from her final stop at Shaffers Crossing. Much like October 24, 1959, she is back in Roanoke after a final trip and headed for retirement again.

Norfolk and Western Railway
STREAMLINED PASSENGER LOCOMOTIVE Class J
Type 4-8-4
Road Numbers 600 to 613, inc.
Built at Roanoke Shops

Cylinders	27” dia. x 32” stroke
Valves	14” dia. piston
Tractive Effort	80,000 lbs.
Superheater	Type “E”
Stoker	Standard “HT”

BOILER

Type—Radial Stay

Diameter:

Maximum O. D.	102”
First Course O. D.	92”
Working Pressure	300 lbs.
Heating Surface:	
Total Evaporative	5271 sq. ft.
Superheating	2177 sq. ft.

GRATE

Length	12’ 2”
Width	8’ 10%”
Area	107.7 sq. lt.

TUBES

227	3-1/2” O. D.
63	2-1/4” O. D.
Length	19’ 21/2”

DRIVING WHEELS

Diameter Outside	70 in.
Diameter Center	62 in.

JOURNALS

Roller Bearings on all Engine and Tender Axles.

WHEEL BASE

Driving, rigid	18’ 9”
Engine	47’ 3-1/2”
Tender	35’ 10”
Total Engine & Tender	95’ 4-3/4”

COUPLED LENGTH

Engine	61’ 2-1/4”
Tender	47’ 11-3/4”
Total Eng. & Tender	109’ 2-1/4”

WEIGHTS (In Working Order)

On Driving Wheels	288,000 lbs.
On Front Truck	101,600 lbs.
On Trailing Truck	104,400 lbs.
Total Engine	494,000 lbs.
Tender (Loaded)	378,600 lbs.
Total—Eng. & Tender	872,600 lbs.

TENDER

Water Capacity	20,000 gallons
Coal Capacity	35 tons

Equipped with Needle Roller Bearings on Valve Gear, Timken Light Weight Reciprocating Parts and Roller Bearings on Crank Pins and Crosshead Pins.

Cutting a fine sight with Train No. 25, the classic *Powhatan Arrow,* the 607 is heading west on its daylight journey to Cincinnati. The Class J has three coaches, a diner and round-end observation car trailing at Pearisburg, Va. in this 1952 scene. *Norfolk and Western photo*

Roanoke Machine Works No. 117, was the first locomotive outshopped by what would become the Roanoke Shops of the Norfolk and Western Railway. *Courtesy Norfolk Southern Corporation*

Introduction

Pretzel-like swirls of decorative, gingerbready scrollwork from the south facade of Hotel Roanoke reflected from the deep pools created in the highly-polished, streamlined cowling and tender of Norfolk & Western class J 4-8-4 No. 611 as she promenaded past Jefferson Street Crossing. After her public display, she was slowly chuffing her way west to the big Shaffers Crossing Roundhouse and Terminal, to take her rightful place on N&W's renowned roster.

PART 1

Dating from 1884, when 86-ton Class I 2-8-0 No. 117 was turned out by the predecessor Roanoke Machine Works (the fledging city's first industry) at a cost of a little more than $12,000, Norfolk and Western built steam locomotives at Roanoke, Va. until 1953. In all, a total of 447 engines were constructed at what was known locally as the "East End Shops," after the RMW was acquired outright in 1897.

Over the years, the railroad became an American and global leader in motive power design, innovating several industry firsts. Much has been chronicled about N&W's steam accomplishments, but this work is intended to be only a simple history of one special locomotive. However, it is important to have some understandable background.

N&W's Motive Power Team

As steam technology developed, those occupying important positions on the N&W team included: R. G. Henley, appointed superintendent of Motive Power in 1928; Mechanical Engineer J. A. Pilcher who retired in 1938

You could tell that she was special, even on that sunny, May 29, 1950 when she first saw the naked light of day. As the representative (first of three) locomotive of the third and final batch of her sisters constructed, Norfolk & Western had staged Class J 4-8-4 611 – freshly-outshopped with the paint still pungent – for public display at the west end of Track 4 of the Roanoke Passenger Station. I mean look how lovingly the shopmen are going over her streamlined flanks, polishing them with tenderness to a high sheen! And see how the whispy, cumulus clouds bounced off her bullet-nosed shroud. Instead of a broadside builders portrait, what a fitting photograph that fairly screams her company heritage! Besides, "Six-Eleven" had a nice ring to it and would prove to be a lucky number, indeed. Some 65-years-ago, it was already evident that she was destined to be...*3 TIMES A LADY.*

H.W. Reynolds

and was succeeded by H. W. Reynolds; and Chief Draftsman Gurdon P. McGavock. Working under McGavock's watchful gaze were such stalwarts as C. H. Faris, V.C. Glaze, George Kimmerling, E. C. Razenor and A.K. Simmons (All told, there were about 15 in the drawing office.).

Others who played important roles were Engineer of Tests H. W. Coddington who was succeeded by I.N. Mosely and then R.M. Pilcher (son of J. A. Pilcher). This office was primarily occupied with the operation of the Dynamometer Car (Roanoke designed and built in 1919), in addition to research responsibilities.

Robert M. Pilcher

Many of the above-mentioned were Mechanical and Engineering graduates of the Virginia Polytechnic Institute. Others made their way up through the ranks of the shop hierarchy, progressing from apprentice to machinist, from roundhouse foreman to master mechanic.

In 1952, Mr. Henley retired and was succeeded by Clarence E. Pond, his long-time assistant. Pond shepherded the railroad through its final days of steam glory, when it basked in the limelight of being the last American stronghold of these charismatic locomotives.

Clarence E. Pond

The reputation of these gentlemen preceded themselves. A representative of the famed Lima Locomotive Works was once asked whether one of their exemplary commercial designs had been offered to the N&W. "I don't believe we did," replied the salesman. "After all, what could we tell *them* about building a steam locomotive?"

The Modern Coal-Burning Steam Locomotive

Commencing with the Class A 2-6-6-4 in 1936, Norfolk & Western developed what it called the "Modern, Coal-Burning Steam Locomotive (i.e., a unit with a high-capacity boiler, roller bearings on all engine and tender wheels, one piece, cast steel frame, improved counterbalancing and complete mechanical and pressure lubrication)." Culminating in three classes—the A, J and Y6—these "Finest Steam Locomotives Ever Built" gave the road a versatile trio that could efficiently handle virtually any assignment for which they were called upon.

The "A" could easily move long coal and time freight (scheduled merchandise) trains over the flatter portions of the N&W. It could also pinch hit on passenger or troop trains at better that 70 miles an hour.

Class "J" locomotives, of which No. 611 is the surviving representative, were a stable of high-speed, streamlined passenger power. They were capable of clipping off 15,000 miles a month at occasional speeds of up to 110 miles an hour!

In the "Y6," what with an even 80 locomotives in its 2-8-8-2 configuration, was the plodding workhorse of the railroad. It packed a heavy punch for the mountain grades and curvature of the coalfield branch lines.

Together, these three road types accounted for 84 percent of the railways passenger miles and more than 90 percent of freight-ton-miles. This was truly amazing, considering that they comprised little more than a third of the locomotives rostered!

Now that Y6a No. 2156 has arrived back in Roanoke on a five-year lease from the National Museum of Transport in St. Louis, the Virginia Museum of Transportation, with Nos. 1218 and 611 already in fold, is proud to display all three types of these prominent Roanoke-built engines.

While none survive, N&W also perfected a steam switch engine, bought from and then duplicated from a Baldwin-built, Chesapeake & Ohio design, which itself was copied from a 1918 USRA standard. When S1a 0-8-0 No. 244 rolled out of Roanoke Shops in December of 1953, it had the honor of being the last reciprocating steam locomotive constructed in the United States!

From the Pocahontas Division section foreman in remote Roderfield, W.Va. to the steam champion in the president's chair, *"our engines"* is the way proud N&W employees referred to *their* locomotives. So and with an elementary understanding of N&W's basic steam designs and philosophy, let's turn our focus to a treatise of its most fabled product: the Class J 4-8-4.

The Coming Of The Class J

With World War II clouds brewing over Europe and having the Class A as a dual-purpose cover girl for heavy passenger/time freight and the Y6 as a dependable scrub

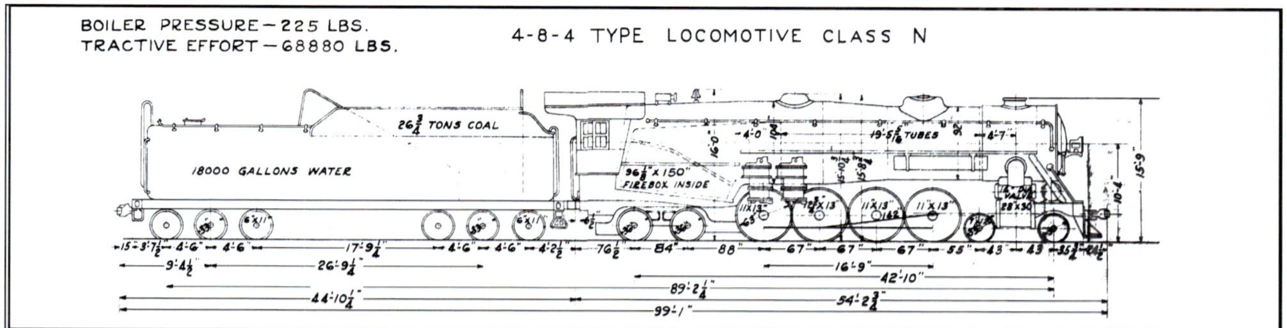

WEIGHTS IN WORKING ORDER	FIREBOX — LENGTH 150"	JOURNALS, DRIVING, MAIN 12¾" X 13"	WEIGHT OF LOCO. PER CYL. H.P.	TENDER CAPACITY, WATER — 18000 GALS.
ON DRIVERS	FIREBOX — WIDTH 96⅛"	JOURNALS, DRIVING, OTHERS 11" X 13"		TENDER CAPACITY, COAL — 26¾ TONS
ON ENGINE TRUCK	HEATING SURFACE	JOURNALS, ENGINE TRUCK 7" X 12"	CRANK PINS, MAIN — 10" X 6" & 9" X 10"	TENDER TRUCK FRAME — CAST STEEL
ON TRAILING TRUCK	TUBES & FLUES 4424 SQ. FT.	JOURNALS, TRAILING TRUCK	CRANK PINS, INTERMEDIATE — 7" X 5 5/16"	TENDER TRUCK BOLSTER — CAST STEEL
ENGINE, TOTAL	FIREBOX 258 SQ. FT.	VALVE 14" PISTON	CRANK PINS, FRONT & BACK — 7" X 4 5/16"	BOILER PRESSURE — 225 LBS.
TENDER	ARCH TUBES 43 SQ. FT.	TRAVEL — 8¾" LAP — 2"		SAFETY VALVES — 3 - ASHTON 3½"
ENGINE AND TENDER	COMBUSTION CHAMBER 109 SQ. FT.	LEAD — ¼" EXHAUST CLEARANCE — 1/16"	VALVE MOTION BAKER	SUPERHEATER — SCHMIDT - TYPE A
TUBES — NO. 245 DIA. 2¼"	TOTAL HEATING SURFACE 4834 SQ. FT.	STEAM PORTS — 1 11/16" X 33"	REVERSE GEAR — FRANKLIN PRECISION POWER	STOKER — DUPLEX
FLUES — NO. 58 DIA. 5½"	SUPERHEATER 1380 SQ. FT.	EXHAUST PORTS — 1 11/16" X 33"	FEED WATER HEATER — WORTHINGTON NO. 4 BL	CYLINDERS — 28" X 30"
TUBES AND FLUES — LENGTH 19'-5 5/16"	GRATE AREA 100.1 SQ. FT.	MAXIMUM CUT OFF 75 %		BUILDERS —
ARCH TUBES — NO. 5 O.D. 3"		CYLINDER HORSEPOWER (COLE) 3173	TENDER FRAME — CAST STEEL	YEAR BUILT —
SUPERHEATER TUBES (UNITS) — O.D. 1½"		FACTOR OF ADHESION	JOURNALS, TENDER — 6" X 11"	

woman to clean up coal and other traffic on the mountainous regions, the only thing the railroad needed as the 1940s approached was a modern engine for pure passenger power. The desire for such a locomotive became increasing critical as the railway's passenger traffic grew five-fold shortly after 1939, as a result of more military traffic and the fact that aging E-Class 4-6-2 Pacifics and K 4-8-2 Mountains could no longer handle the volume alone.

Originally, the design team considered a Class N 4-8-4 with 63-inch drivers as a passenger engine, but rejected it. Remembering the problems encountered with the K-3 4-8-2s (a not-very-successful design of 1926) while employing the same driver size. What they wanted was something perhaps a little special, maybe even a "movie star," of the same wheel arrangement with 70-inch drivers that could be counterbalanced against wheel slippage, even for theoretic speeds of up to 140 mile-per-hour!

Rather than turn to some fancy New York design firm, the Mechanical Department went

Frank C. Noel's final design featured a Tuscan red band trimmed in gold with gold lettering and numbers as suggested by his wife. *Both: Norfolk and Western*

The "Big Three," from left Class J No. 604, Class Y6 No. 2147 and Class A No. 1212. Class A locomotives were numbered 1200-1242, while the Y6 classification was reserved for the 2120-2200. But N&W had an even 200 of the 2-8-8-2 types on its roster. This pose was recorded several different times, with different locomotives over the years. *Norfolk and Western photo*

Clockwise from center top: The 600's nose is a fine example of the tinsmith's art as this view shows the different pieces welded together.

The tender is taking shape in the Tank Shop.

Bottom right: Tinner Richard Lisle (right) and his assistant are forming up the many pieces of the skyline.

Bottom left: The boiler is hung from end to allow for hammering by the "Bull Riveter." Hanging it vertically allows for reduction in stress on the metal. The boiler is shown in the tall end of the Boiler Shop depicted in the post card and aerial view on pages 6 and 15. *All: Norfolk and Western photos*

The running gear is coming together on the 600 in this overhead view. Of note here are the many lubrication lines. *Norfolk and Western photo*

The Roanoke Shops, also known as "East End Shops" covers 57 acres. It was in this place that 447 steam locomotives were built new, including the last Class One steam locomotive built in North America. *Norfolk and Western photo*

220 points. The system could operate 1,300 miles without replenishment and the roller bearings on the wrist and crank pins could operate 500 miles without greasing!

The 52-foot-long boiler carried an initial working pressure of 275 lbs. and was constructed of 4,437 flexible staybolts and 2,420 rivets. The boiler with its Type E superheater and Worthington 6SA feedwater heater, was capable of evaporating 225 gallons of water-into-steam-per-minute when heated to a maximum by the 107.7 square feet of grate area in the firebox, fed by a Standard HT stoker.

In working order, the Class J locomotive and tender weighed in at 872,600 lbs., with a length of 109 feet, two and one-fourth inches. The rectangular, streamlined tender rode on roller-bearing Commonwealth trucks and had a capacity of 30 tons of coal and 22,000 gallons of water. Since water tanks were more common than coaling stations, these were later modified by increasing the former to 35 tons and sacrificing 2,000 gallons of the latter—still the largest tender ever built by N&W.

Tractive effort for four of the five locomotives was 73,300 lbs. The exception was the 602, which was fitted (until December 1945) with a Franklin Type E booster, giving it a rating of 85,800 lbs. and making it the most powerful locomotive of its type in the world!

Getting To Work

The No. 600 rolled out of Roanoke shops on Monday afternoon, October 20, 1941. After testing and some exhibitions at key points along the system, it was assigned to Shaffers Crossing for pool service out of Roanoke, mostly to Williamson, W.Va. and on through Southern trains to Bristol, Va. The four other new passenger locomotives soon followed suit (the 604 was completed in January 1942), shortly just before and after Pearl Harbor.

As a complement to and just prior to the introduction of the spiffy new locomotives, 15 streamlined Tuscan coaches were ordered from Pullman Standard and placed in service on the *Cavalier* and *Pocahontas* and 35 used Pullman chair cars were reconditioned as PN coaches at Roanoke Shops. Southern's silvery *Tennessean* was also exhibited, having made its debut in regular service on May 17, 1941 and modern, rebuilt "luxury" cars were put in service on SR's *New York, Chattanooga & New Orleans Limited* (later renamed *The Pelican*) that same month.

The Js were ideal for quick starts in getting away from stations and accelerating out of speed restrictions on tight curves on the Radford and Pocahontas Divisions. Occasionally, the first five would venture out on the Norfolk Division, where they could attain their ideal

A group of shopmen are working at the final fitting up of the streamlining on the 600. Within a matter of days, she will be ready for a trip to the Paint Shop. October 1941. *Norfolk and Western photo*

operating speed of 80 m.p.h on the flatlands east of Lynchburg. They were seldom used on the Shenandoah Division or Scioto Columbus District because they would have been a waste of power with only 5/6-car trains.

Six More For The War

By 1945, N&W passenger traffic had mushroomed to 374 percent, as compared to pre-war figures. To meet that ever-increasing demand, the railway authorized construction of six more of the 4-8-4s, Nos. 605-610, at Roanoke Shops at a cost of $168,550 each. Production commenced in October 1943 and was complete by December of that year.

Classified as J1, they were essentially the same locomotive as their earlier sisters, except the streamlined cowling was omitted and the rods were not equipped with roller bearings. Citing material shortages, these changes were dictated by the War Production Board.

Another difference was the switch to the Alligator steel crosshead of Timken design, with aluminum, babbit-lined bearing surfaces that clamped around the piston rods. This change was made because the previous application was difficult to maintain.

Although the lack of streamlining on the J1s caused no problems, lending to a rather brutish, handsome appearance (co-author Hensley especially realized this when he saw the 611 undressed at Spencer), the absence of the roller bearings was another story. Despite having bushings in them so that they could later be converted, the immense, single I-section side rods were fashioned out of ordinary carbon steel and were broad in width (but narrow in thickness), in an effort to achieve as much counterbalancing as possible, despite their heavy weight. At speeds above 50 miles-per-hour, the centrifugal force against the plain, floating bronze bearings wouldn't let the lubricant feed; thus, shutting off the grease supply and resulting in overheated main rod bearings.

Assistant Superintendent of Motive Power C.E. Pond personally made several trips to Washington to plead the railway's case for roller bearings on the J1 before the WPB, all to no avail. To correct the problem, several substitute bearings were tried without success.

Draftsman V.C. Glaze recalled that, "We'd try out a bearing and run the locomotive sometimes not farther than 20 miles down the road before having a breakdown and we'd have to bring her back in and try something else. We wasted more brass in trying to get a fric-

The 600 on tour, October 1941. Heading west to Williamson, W.Va. with the new coaches in this view from the Kentucky side (above). Arrival in Williamson brought out most of the town (below). Final display stop was Norfolk Terminal Station (bottom left). *All: Norfolk and Western photos*

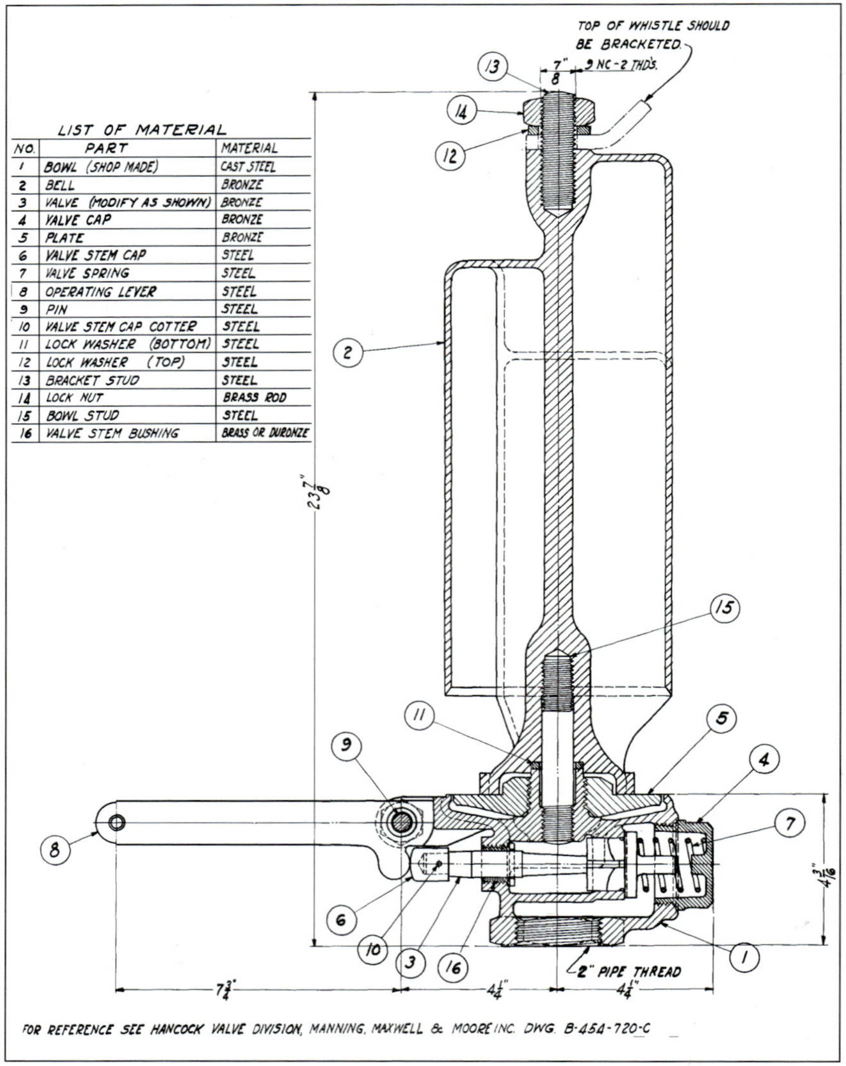

Left: The human interest element was always present. This view of curious, five-year-old Tommy Gray, son of Alexander Gray, helper machinist in the Yard Department, Roanoke Shops, adorned the February 1956 cover of the *Norfolk and Western Magazine*. However, the real story behind this photo was that it originated with a snapshot taken by Mary Margaret Slicer, who was an avid photographer and N&W company nurse. She sent the photo to the Public Relations & Advertising Dept., which loved the idea and assigned a company photographer to duplicate her effort. *Norfolk and Western photo*

Above: The voice of the Class J; the Hancock Inspirator whistle is responsible for the deep-toned sound so familiar to those who listen to the Link recordings or follow the 611. *Norfolk and Western Historical Society*

Posed at what would become a favorite portrait spot, just west of Shaffers Crossing, the 600 gleams in the afternoon sunlight on the freshly laid and surfaced track. *Norfolk and Western photo*

It is hard to imagine in the days before Photoshop, but retouching was a fairly common procedure. It was considerably more difficult and required a totally different skill set than today's electronic world. But the N&W wanted a view of the new streamliners to be used for publicity. Seldom correctly identified is this view of the 600 on No. 4, the *Pocahontas*, below. The top left is the commonly seen retouched view of the train with Virginia's Peaks of Otter in the background, telegraph lines removed, the foliage on the bank cleaned up and train streamlined. The bottom right shot from March 24, 1942 is the un-retouched original image. Note the heavyweight cars and cinders alongside the embankment. You can even see one of the two cameras almost opposite the telegraph pole. The retouched art was used extensively almost up to the end of steam; a color painting variation was used on playing cards that were sold onboard the passenger trains. *All: Norfolk and Western*

J1 No. 606 shows off her brutish appearance just west of Shaffers Crossing. The massive side rods of carbon steel with conventional bearings and subsequent problems plagued the J1s for most of their service time. *Norfolk and Western photo*

tion bearing that would hold up on the J1 than if we'd been allowed to use roller bearings in the first place!"

As the war came to a close, all the J1s were shopped with streamlining, tandem lightweight rods and roller bearings applied by 1945 and reclassified as regular Js. Also, the booster was removed from the 602 in 1945 and the boiler pressure raised to 300 p.s.i. on all members of the class, thus giving the J a tractive effort of 80,000 lbs. and making it the most powerful class of Northern-type locomotives outright.

Pronouncing its philosophy in steam design and servicing as sound after Germany and Japan surrendered, the N&W was content from then on to fiddle with minute refinements or upgradings to each class of locomotive as they were built or brought in for shopping. An example of this was the modernizing (through 1947) of 22 K2 and K2a class 4-8-2s to resemble the racy Js. However, they were easily discernible by the oversized, Worthington 4BL, feedwater heater on the left or fireman's side. Known as a "Red K" or "J Junior" by the engine crews, they brought some class to secondary connecting (Columbus District, plus Shenandoah Valley and Winston-Salem lines) trains and main line locals, particularly through the coalfields.

Three More Makes Fourteen

Making a final push for passenger service, Norfolk and Western had inaugurated the crack *Powhatan Arrow* on a 16-hour, Norfolk-Cincinnati schedule in 1946. As if that were not effort enough, the "Arrow" was updated with "Fine New Feathers" in 1949, featuring a fabled "boat tail" tavern-lounge-observation car on the rear.

Thus – and just in time for Korea, three more of the Class Js, Nos. 611-613, were constructed at a cost of $251,544 apiece in 1950. Incorporating the refinements made to earlier sister engines, plus moving the bell up underneath the front end, these final locomotives completed the class at a count of 14.

The only major change brought out on the last batch was the use of one-piece, cast steel wheels on the lead or pony truck, as it was more economical than the small, spoked-

An excellent view of the tandem rods on newly modernized Class J No. 610. *Timken photo*

with-tires wheel used on previous members of the class as built. These new pony wheels were applied to Nos. 600-610 during subsequent shoppings during the early 'Fifties.

Also, in 1952 (as verified by an N&WHS drawing) and at the "suggestion" of Timken, according to C.E. Pond, the tandem rods were discontinued and a simplified arrangement applied. This new standard, requiring less maintenance, was retrofitted to the 600, 604, 607 and 611 during scheduled major shoppings. It called for a single intermediate rod, with the remaining back rod inside and closer to the wheel face. Yet today, debate rages as to where these new rods were actually stamped out. It is known that Timken supplied Roanoke with their specialty steel billets, but the serial numbers on 611 are similar to those manufactured for other railroads at Canton, Ohio (specifically, Ross Rowland's C&O "Greenbrier" 4-8-4 No. 614 comes to mind).

Performance Plus and Speed To Burn

Beneath that clean, streamlined cowling, the Class J possessed a brawn and swift gait that belied its pristine appearance.

In a 1945 dynamometer-car test up the Allegheny grade between Elliston and Christianburg, Va., No. 604 developed a maximum drawbar horsepower of 5,027 while pulling a 16-car, 1,065-ton train!

No one ever found out how fast they could actually go. The No. 610 was borrowed by the Pennsylvania Railroad for road tests between Crestline, Ohio and Chicago in 1944 and was clocked at speeds of more than a hundred-miles-an-hour – hitting a maximum of 110 – for mile after mile on level, tangent track. As drivers traveled in a blur at 18-feet, 4-inches per revolution (288 per mile), N&W Vice-President C.P. Blair remembered hitting 112 with a J on the Cincinnati District near Wil-

The K2/K2a Class was streamlined after World War II to look similar to the Class J, giving a streamlined, modern look to secondary trains and on lines that were normally restricted for the Class J. N&W "dieselized" its servicing, bringing faster handling of locomotives in terminals, making power more efficient and readily available. K2a No. 134 is shown both inside and outside of the new (1947) facilities at Williamson, W. Va.
Both Norfolk and Western photos

Naturally, the Js were used in company promotional material. Top left: A ticket envelope reminded the public that the N&W offered three fast trains. Top right: A hand-tinted view of the 610 rushing westbound at Glenvar, Va. with what was actually the *Tennessean*, but has been painted to be a Tuscan train, was on the cover of this luncheon menu from the dining cars. Bottom right: Of course, the rear of the *Powhatan Arrow* was also shown with its boat-tailed observation car posed at Dry Branch, Va. Below: One of several cartoons over time printed in the *Norfolk and Western Magazine*, February 1944.

"He started firing on the streamliner this morning . . . and look what happened!"

Lionel's model of the Norfolk and Western Class "J" streamlined steam passenger locomotive takes on a big job as it attempts to pull its huge counterpart while N. & W. President R. H. Smith and veteran Class "J" locomotive operating Engineer D. G. Garrett and his two grandchildren look on. The grandchildren are Barbara Tate, 9, and Steve Garrett, 8. The N. & W. locomotive 610 has traveled 2,100,000 miles. (Photograph from Bernard Relin & Associates, Inc., for the Lionel Corporation)

NORFOLK AND WESTERN MAGAZINE

The Class J is an enduring model subject, leading off with the Lionel version, left, in 1957. However, the first model of the J was one made by the shops to show off before the 600 was actually completed. It is believed that several of those were built, at least one came to be on display in the Roanoke Passenger Station (above, 1957), with Hotel Roanoke visible through the large windows. A larger, live steam model was built by Machinist W. E. Cowling of Bluefield; the company frequently borrowed it for a variety of promotional events, including parade floats, as shown below in the "Diamond Jubilee," Roanoke 75th Birthday celebration in September 1957.

liamsburg, Ohio when he was superintendent of the Scioto Division and Engineer of Tests Robert M. Pilcher recalled riding across the Dismal Swamp at 115 on the Norfolk Division!

The J was a railroad operating man's dream. In a 1982 interview with co-author Hensley, C. E. Pond thought that they were the best engines ever designed by N&W, because of their being "tops in refinement, serviceability, performance and availability."

Although they never picked up on the design, perhaps because of parental vanity (up until the mega-mergers, the Pennsylvania had a controlling interest in Norfolk & Western). PRR's Mechanical Engineer told G.P. McGavock after testing the J that "it was the best engine that has ever been designed in the country," continuing that "there are a number of locomotives similar to it being built, but when they are completed, it will still be the best."

They were tops with the crews, too. About them Norfolk Division engineer J. L. Walrond remembered simply "they were the best."

"They fired easy and rode good," Walrond said. When they were working hard up on Blue Ridge, "You could see the fire raise every time that engine would exhaust," he continued. "I've seen 'em down on the 'dollar mark' (100 m.p.h.) through 'the bottom' at Appomattox 'a many 'a time!"

Climbing the Eastern Continental Divide into Christiansburg, F.T. Nichols of O.Winston Link, Abingdon Branch bootleg whistle fame, recalled how they performed when he was a Radford Division passenger engineer. "You take that J with 16 cars and it would really walk that grade," he said. "And I mean *WALK* that Allegheny Mountain!," he repeated for emphasis.

But, perhaps it was a doting American public that most admired the Class J. A picture of one hung in a restaurant as far away as Kansas and Lionel bestowed the honor of copying it as a prototype for one of its O-gauge models in 1957. Even *TOOTLE*, the prize student of the Lower Trainswitch School for Locomotives in Gertrude Crampton's 1945 children's storybook classic, was said to be "the finest baby I've seen since old 600!"

The gloomy weather is appropriate as a future steam star is leading an eastbound *Pocahontas* through Glenvar, Va. At this moment in time, 611 is just one of the 14 Class Js, but in a short while, would become the sole survivor. Soon, diesels will be doing this task, but with a lot less style. *Norfolk and Western photo*

Countering The Diesel

Other railroads, starting with the first successful use on the Central of New Jersey in 1925, turned to the internal combustion means of motive power. By 1950, more than half the wheels of railroad propulsion would be turned by these new engines. In 1953, Southern Railway System – N&W's eventual, Norfolk Southern merger partner – would become the nation's largest, all-diesel line.

But, the N&W held to steam longer than the rest. Spanning the Virginias and reaching into Ohio, it had a fierce loyalty to the on-line originated Pocahontas coal traffic from which it derived the biggest part of its revenue. By 1955, it was the last Class 1 railroad in the country that had yet to employ a single diesel unit.

Another reason was that N&W "dieselized in theory." Once they had perfected their storied designs, management turned its attention to construction of modernized maintenance and servicing facilities. These, along with other concepts such as the "canteen" auxiliary water tenders (which allowed the retirement of expensive, intermediate tanks between division points), resulted in a higher availability and less time in getting over the line.

Stubborn to the end, the railroad tried a last ditch effort to thwart the diesel, when it partnered with Baldwin-Lima-Hamilton and Babcock & Wilcox in the 1954 construction of an experimental steam-turbine-electric dubbed the *"Jawn Henry."* Troublesome during testing, TE-1 No. 2300 nearly had all the bugs worked out when a decision to construct five copies was turned down by the Operating Committee, prior to an N&W Board of Directors

Only weeks before this would have been a Class J leading the *Pocahontas*, but for now, leased Atlantic Coast Line units have been hastily relettered as "Norfolk and Western." The water plugs are no longer needed for this motive power. *All: Norfolk and Western photos*

By December 1958, N&W's own passenger units, GP-9s 500-521, had begun to arrive on the property and the leased units were being returned to their owners in time for their own winter passenger rush to Florida. The 519 and 511 have the *Pocahontas* posed at Montvale, Va. this chilly day. The cut on the mountains in the distance is the famed Blue Ridge Parkway. A clue about the demise of passenger service is also shown, as three mail cars and a Railway Post Office lead the train and two box express and an office car bring up the rear. Not hard to see where the train's money was being made.

A large group of EMD representatives are posed with N&W officials as the completion of dieselization is announced. Like a proud new father, a beaming Stuart Saunders (right) is contrasted with a rather unhappy looking R. H. Smith (third from right). Clarence Pond is fifth from the left.

meeting. When Vice-President and General Counsel Stuart T. Saunders (who had been listening to the PRR representatives seated at the table) announced that he would oppose the recommendation, President (since 1946) R. H. "Race Horse" Smith "got so mad that he kicked a spittoon clear across the room!"

Out Of Time

Still, by the mid-1950s, N&W's Class J-drawn "Pow'tan" was being touted in trade publications as the "nation's last great steam train." Those on the West Coast may argue on behalf of Southern Pacific's flashy, orange *Daylights*, but, especially on an Autumn day, Nos. 25 and 26 surely were a thing of beauty to behold as they scorched the broad Ohio River bottom lands, rustling the sun-bleached sheaths of corn. After winding through the lush New River Valley, they bolted across Southside and Tidewater Virginia, where they averaged 65 or better, start to stop!

But, the handwriting was on the wall. It was clear that rising labor costs and an increasingly arduous, as well as expensive, effort in locating component parts would soon bring an end to this *Camelot* story of steam.

After being featured as the "story of the year" in a *Trains* magazine "Faith In Steam" article by noted Editor David P. Morgan in the November 1954 issue, Norfolk and Western succumbed to internal combustion power the next year, when it was announced on May 29 that it was buying eight diesels. In press releases, the official reason given at the time was that they would "free up" bigger steam power for mainline service.

These "invaders" were placed in operation on the Durham District of the Norfolk Division. This was soon followed with an order for 190 more and the Shenandoah Division, Bristol Line of the Radford Division, Clinch Valley Line of the Pocahontas Division and Cincinnati District of the Scioto Division were virtually all-diesel operated by the end of 1957.

Stuart Saunders succeeded Bob Smith, when he retired on March 31, 1958; the former later went on to wreck the Penn Central. "He had no emotion for steam whatsoever," observed W. T. "Bill" Ross who retired as N&W's general superintendent of Transportation

Two months later in June, one of his first official actions was to announce that the N&W had placed a $50M order with the Electro-Motive Division of General Motors for 268 GP-9 diesel units. The action "will enable us to dieselize competitively our operations at present traffic levels," Saunders said.

Also, in July of that year, the Js were taken out of passenger service and their bullet-nosed front portrait was symbolically –if not unceremoniously –removed from the cover of the public timetable. This became evident in the October 26, 1958 circular.

All 14 Js received a brakeman's cupola or "monkey box," as the crews teasingly called them and they were placed in standby passenger or freight service: Usually, they were a little too much locomotive for the job, but they were used on Norfolk Division locals and Scioto Division time freights between Williamson and Columbus.

In the "Facts" section of the September 1958 *Norfolk and Western Magazine*, it was reported that the 4-8-4s had operated approximately 30,450,000 miles since they were built at Roanoke Shops –equivalent to the distance that would be involved in about 64 trips from the earth to the moon and back!

Norfolk & Western steam would last a little longer – until the wee hours of May 7, 1960 to be exact. The end came at Williamson, where a handful of Y6 locomotives for mine run shifter service and S1a yard switchers had retreated.

But, all the sleek Class Js had been retired and consigned to scrap by mid-1959... save one. Its number: 611!

••••

Noted *The Virginian Railway* author H. Reid also spent a good bit of time recording the N&W. Reid was one of the masters of composition as far as railroad photographers go. He captured the 602 with 10 cars rolling at speed on the Norfolk Division.

The 613 is taking water at Disputanta, Va., while working out her remaining flue time in local freight service on the Norfolk Division, March 1959. The six cars and caboose were light work for the big 4-8-4.
H Reid photo

H. Reid captured the 608 working a local freight near Zuni, Va., January 18, 1959.

Sister 612 with canteen is powering Time Freight No. 94 east and just passing KX Tower at Kenova, W. Va. *Bruce R. Meyer photo*

The first class J, No. 600, is also in local service with the pulpwood piled almost as high as the cab in this January 1959, Norfolk Division scene. *H. Reid photo*

Head of the demoted class: The 611 is at the front of the group of four Js stored at Bluefield in January 1959. Steam is still working, but alongside the Alco invader at extreme left. The other three Class Js will be retired and sold for scrap in May. *Norfolk and Western photo*

Many N&W steam locomotives met their demise at the M.D. Friedman scrapyard in Portsmouth, Ohio. From top left: The boiler as shown on this string of locomotives at Star Yards, is the first and easiest step in cutting them up. Behind the S1a, the second locomotive in the line is Class J, No. 600. Bottom right, a T6 switcher is taking this J to be weighed prior to the first cuts. Bottom left, the ivory hunters have already been at work on the 601, as the tear-drop class lights and builders and cylinder class plates are gone. *John P. Killoran photos*

Storming out of Roanoke, Class J, No. 611 is leading eastbound Train No. 4, the *Pocahontas*. Classic scenes like this were repeated day in and day out, year after year. One would wonder if any of the designers would look out of the window of the Motive Power Building (at right) with pride when one blasted away from the station?

Our Heroine's History
What's In The Life Of A Locomotive?

While information grows more sketchy over the passage of time, much can be gleaned from publications, government reports, working timetables, old newspaper microfilms and taped oral history interviews with retired railroaders (most of them now deceased). Motive Power and N&W Division records are another source. Much was evidently thrown out from other headquarters, but particularly good data was saved off the Scioto Division, headquartered at Portsmouth.

In chronological order, we have been able to piece together much about No. 611.

From pages 505-506 of his *Rails Remembered, Volume 3*, Louie Newton reported that he was working at the Shaffers Crossing coal wharf on May 29, 1950, under the tutelage of Charlie Walton, a black laborer. Sometime in the early afternoon, someone in the area called out, "Look, there's that new 'J' engine!" On its Tuscan red skirt it carried in gold enamel the number "611." When its builders plate was examined at close range a few minutes later, it was ascertained she was the 388th new locomotive turned out by Roanoke Shops.

A month later, on June 29, she paid a visit to the drop pit at Shaffers for her first monthly inspection. Assigned to Machinist Bill Gulledge and his helper, Newton and his two colleagues "removed the right cylinder head, inspected the right piston packing and found it to be worn 1/8 of an inch." The testing, inspection and necessary work were completed in a rapid manner: It arrived at the roundhouse about 7:00 a.m. and left at 11:25, to take out No. 9, a Bristol local, an hour later.

Later that year in what was likely her first skirmish and while in charge of Train No. 4, the *Pocahontas*, No. 611 struck a 1950 Ford at the Gennetts road crossing (near Franklin Furnace, Ohio) on December 2. An examination of the *Ironton Tribune* revealed no further pertinent information.

In the early 1950s on the Norfolk Division, a vehicle drove into the side of a train powered by No. 611.

Then, the year 1952 proved to be an eventful one for our favorite locomotive.

On May 14 at 3:38 a.m. while handling westbound No. 3, she "creamed" a loaded tractor-trailer belonging to the Telling Ice Cream Co. (of Columbus) at South Point, Ohio – not far across the Ohio River Bridge from a scheduled stop at Kenova, W.Va. Headed south on U.S. 52 to make a delivery to Huntington, driver Fred L. Arnott of Lockbourne, Ohio narrowly escaped death after he had dozed off and was awakened by the wig-wag crossing signals heralding the approach of eastbound Time Freight No. 86. He applied his brakes, skidded about 100 ft. and veered left up the N&W Westbound main, directly in the path of the passenger train. He barely had time to escape before the 611, in charge of Engineer C. Johnson of Portsmouth and Conductor E. Bowdle of Cincinnati, reduced the rig to splinters. An investigation was conducted by the Ohio Highway Patrol and N&W Yardmaster Bob Stewart and the big 4-8-4 and her train proceeded west, leaving behind a melting, gooey mess.

611s Highest Honor

Other than being a survivor, the high point of her working life probably came on September 24, 1952 (co-author Hensley's third birthday!), when she handled the special of General Dwight. D. Eisenhower, Republican candidate for President of the United States. Teamed as the trailing locomotive of a doubleheader with sister No. 613, she left Columbus Union Station at 6:30 that morning, bound for Kenova. The 18-car train was delivered to the N&W by the New York Central and would be turned over to the B&O, bound for Wheeling, W.Va., up the latter's Ohio River Line.

After a stop at Chillicothe, "Ike" arrived at Portsmouth (on time) for a 9 a.m. speech. Another 15-minute pause was made at Ironton, before the campaign contingent was turned over to the foreign railroad at 10:30. We wonder what The General thought of B&O's rickety track, after the smooth ride he had over the well-maintained Norfolk & Western?

The train continued over the B&O, with the same equipment staged at Baltimore for a September 25 nightime departure to Washington, D.C, thence over the Southern via Charlotte to Winston-Salem. There, the train was turned back over to the N&W at 11:40 a.m. on September 26.

After a speech before a crowd of 10,000 at the Union Station (currently being renovated) esplanade, the special headed up the "Punkin Vine" and on to Petersburg. This time, Class J Nos. 600 and 604 did the honors; fittingly –and perhaps as a harbinger of things to come –the train departed at "12:18" p.m. Class Y6b 2-8-8-2 No. 2176 followed along "lite" as a precaution and N&W diners Nos. 493 and 494 had been cut in during the stop.

Doubleheaded Class Js were certainly not a common occurrence, but the railroad did not want to take a chance with any presidential candidates, so two Js power the Eisenhower special into Portsmouth, Ohio on September 24, 1952. The 613 leads 611 into town, trailing mostly New York Central equipment. You can bet there are more than a few of the "I Like Ike" buttons in that crowd as the popular General speaks to "Peerless City" citizens from the rear platform. *Left, Bill Harmon Collection. Bottom Norfolk and Western photo*

Tim Hensley Collection

Throngs greeted the train along its way up the Shenandoah Division and impromptu orders were issued to slow down to seven-miles-an hour through Martinsville and Bassett, where Eisenhower and wife Mamie waved from the rear platform. The special arrived opposite South Roanoke Park, where a crowd of 7,000 awaited to hear the D-Day Commander speak, while fresh Js 612 and 613 were exchanged to handle the dash (Stops were only made at Lynchburg and Crewe.) over the Norfolk Division for delivery to the Atlantic Coast Line at Petersburg.

Applicable division road foremen rode the locomotives during the entire time the special was on the N&W. Ready to troubleshoot general and division superintendents were also aboard.

Moving ahead to 1953, the No. 611 would have three crossing incidents to deal with, all on the Scioto Division and one of them significant.

On the afternoon of Saturday, June 20, with No. 15, the *Cavalier*, she hit a bread truck on the U.S. 52 crossing at Chattaroy, W. Va. The *Williamson Daily News* had already been put to bed and did not publish on Sunday; it was not until the Wednesday, June 24 edition that it reported that W.R. Viers, a West Williamson driver for Feutchenberger Bakery, was scheduled to have shoulder surgery that morning for injuries sustained in the collision. Apparently, he had stopped for an eastbound train and then pulled into the path of the big Class J, coming opposite on the double-track.

About the time Viers was in the recovery room at Williamson Memorial Hospital, our

girl stubbed her toe again, also at South Point. She suffered considerable damage to her front end and this was likely No. 611's most serious incident until the 1956 Cedar, W. Va. wreck. At 11:55 a.m. on June 24, 1953 while pulling No. 26, the eastbound *Powhatan Arrow*, the Class J clipped the rear end of a tractor-trailer, shearing eight feet off its end, opening it like a can opener and dumping merchandise along the railway's right-of-way for 500 feet. Unable to continue, No. 611 received a dented left nose and the fireman's side front steps were crushed in (photo above).

Again, while waiting for an opposite train to pass, driver Charles R. Calvert of Decatur, Ill. pulled onto the crossing and into the path of the speeding passenger train. Jumping from the truck's cab at the last moment after "gunning" it, most of his trailer had cleared before it was hit (see photos page 38).

Stopping nearly a mile down the track, the *"Arrow"* obtained the dispatcher's permission to back up and inspect the scene of the accident. Seeing that No. 611 suffered considerable damage, Class M 4-8-0 No. 457, a superheated Kenova Yard engine, was summoned from across the bridge; luckily, the convenience of the North Kenova Crossover, controlled by KX Tower, expedited the movement.

Arriving to the rescue, yard engineer Doug Booth, brakeman Walker Johnson & crew traded out their old "Mollie" and trundled the injured J back to Kenova to lick her wounds. She would require some straightening at the running repair shop, before being sent on to Roanoke for heavier work.

As expected, "High Iron" engineer Pat Duncan and his fireman were not happy with their end of the "deal." By the time he hand-fired No. 26 on into Williamson, "It liked to killed John Moreford," the jovial, mischievous Johnson laughed.

In one last 1953 mishap, No. 611 struck a 1949 Hudson with Train No. 25 at Crum, W.Va. About that incident, only the license number – Ohio A-6925 – is known.

Toward the end of 1954, America's most-famous 4-8-4 hit a 1941 Ford sedan with the westbound *Powhatan Arrow* on the evening of December 7 at McDermott, Ohio. It's almost as if No. 611 had a "thing" for commemorative dates in relevance to WWII, although she never participated in it!

The next-to-last documented automobile incident involving the engine in regular service happened on July 26, 1955. While leading the eastbound *Cavalier*, she clipped a 1952 International truck at Cincinnati's Canal Ridge Rd. She also was found to have a worn spring equalizer about Dec. 18 of that year.

While this may sound like many accidents, they were just part of normal, day-today opera-

611 is departing Crewe, Va with Train 4, *The Pocahontas*, bound for Norfolk, Va. on November 19, 1953 *H. Reid photo*

tions. The Js typically operated flawlessly and rarely did anything unusual occur. They usually went yeoman-like from scheduled classified repairs to scheduled classified repairs, with only minor work performed as needed or discovered during daily and monthly inspections.

It is documented that class repairs to our star locomotive were completed on November 23, 1955. She had accumulated an estimated 30,628 miles since that shopping.

Tragedy on the Tug Fork

Next, came No. 611's darkest hour. And in a way, it might have been her saving grace.

Just after midnight of January 23, 1956 and in a light snow, engineer Walter B. Willard, 62, of Princeton and fireman Ernest W. Hoback, 27, of Bluefield, were hustling Train No. 3, the westbound *Pocahontas*, along the "Tug River" (officially, the Tug Fork of the Big Sandy) and over the coalfield, N&W division of the same name. Back in the 11-car consist (total train length of 996.01 feet), conductor Edward N. Camden of Vinton, Va. (The three N&W passenger trainmen seniority districts operated Norfolk-Roanoke-Williamson-Cincinnati.), assisted by two brakemen, had charge of an express car, one mail car, four coaches, one tavern car, one diner and three sleepers, all in the order named.

Passenger trains were restricted to 50 m.p.h between Davy and Williamson; they also had to comply with several curves which further hindered their speed. By Devon, the last open office, they were running 21 minutes late. Out of Bluefield down 14 minutes, additional delay was encountered on the Welch Loop, unloading mail and waiting for an eastbound extra.

Approaching Cedar, where an especially tight curve demanding a 30 m.p.h passenger restriction loomed, they derailed at what the crew and others involved "estimated" at 40 –10 miles-per-hour over the speed limit! The time was 12:51 a.m.

Running on clear signals, a permanent, 30/25 (the lesser for freight) yellow, curve restriction (warning) sign – 20 inches in diameter – was posted 1,771 feet east of the point of derailment, which was 59.98 miles west of Bluefield and 1,412 feet east of the station shelter. Although engineer Willard made a brake application and closed the throttle about Milepost 449 as he entered the curve to the right, No. 611, the first five cars and the front truck of the sixth, came off the track.

The locomotive overturned, plowed over an embankment toward the Tug River and came to rest on its left side; the tender remained coupled but twisted over on its top. The skirting, running board, part of the valve gear and most of the other appurtenances on the left side of the engine were torn off. It came to rest 305 feet west of the point of accident and 68 feet south of the Westbound track.

Killed was engineer Willard. He was found at the throttle, with his glasses in place and watch still running.

Lucky to have survived was fireman Hoback who could perhaps reconstruct the best version of what actually took place. "My first inkling that something was wrong was when

Top two: What is all the commotion? The 611-led eastbound *Powhatan Arrow* is stopped down the track after hitting the rear of a tractor-trailer at South Point, Ohio on June 24, 1953.

Bottom right: More serious damage was done to 611—two and-a-half years later, as we see her laying on her fireman's side after overturning on Cedar curve the early morning of January 23, 1956. The Railway Post Office car has already been re-railed and sent to Williamson, the mail must go through, you know! Coach 539 is laying on its side with the dim winter light reflecting off the windows. A week later, 539 displays her damage at Roanoke Shops (page 39).
All but aerial view: Norfolk and Western photos. Aerial View: Maurice Kaplan, Huntington Publishing Company.

the engine reared up and started turning over to the left," he said in an interview years later. As the locomotive left the rails, Hoback and Willard looked at one another. "He went up and I went down," the fireman said. "I braced my feet up against the boiler. We vaulted clear off the Westbound, 'tore up Jack' and landed almost in the river!"

An independent N&W investigation would prove him right. Reporting to Interstate Commerce Commission inspectors, General Superintendent (of the Western General Division headquartered at Bluefield) C.P. Blair wrote: "Examination of the high rail on the Westbound track did not indicate that the head of the high rail had been peened by the engine wheels. It is not believed that any of the marks on the tie plates, spike heads, joint bars, or ties were sufficiently heavy to have been made by the engine or tender."

Calculating the overturning speed in that same letter that was drafted for his signature by L.A. Durham, Jr., assistant division engineer, Blair continued: "The estimated height of the center of gravity of our class J locomotive is 77 inches. The estimated height of the center of gravity of the tender with the estimated amount of water at the time of the accident (15,000 gallons) and the estimated amount of coal at the time of the accident (18 tons) is 68 inches.

In making the calculations for overturning speeds, we have assumed that the distance between the contact points of the wheel tread with the rail is 58 inches. The exact location of this contact point is not known. It is certainly wider than the gauge width; therefore, the 58-inch assumption is conservative.

Based on the above assumptions, we calculate the overturning speed for the Class J engine on a 13-degree and 30-minute curve with 4 and ½ inches super-elevation to be 53.6 m.p.h. The overturning speed for the tender on the same curve, with the amount of water and coal estimated at the time of the accident would be 56.5 m.p.h. The safe speed for the engine would be 35.4 m.p.h. And the safe speed for the tender 36.9 m.p.h."

About his wild ride, Hoback who caught the job off the extra list and was likely preoccupied with his fire when it happened, had this to say, as quoted in an Associated Press story that ran in the January 23 edition of the evening *Huntington Advertiser*: "I went over with the engine into the river. I don't know whether there was a hole knocked in the side of the engine or whether it was through a window that I crawled out into the cold water. The coal from the tender began pouring over me before I could crawl out."

An unidentified man helped the dazed Hoback out of the water and over a baggage car to the safety of the riverbank. Dry clothes were provided by a baggageman who also applied a tourniquet to his injured arm.

As for the consist itself, separations occurred at each end of the first four cars. The express car (No. 123) was jackknifed (buckled) over the bank and leaning on its left side, with the front end resting near the rear of the tender. The mail car (No. 94) stopped on its right side approximately parallel to the track. The first coach (No. 539, third car in train) came to rest on it left side; the front end was near the rear end of the second car. Next, the fourth car (No. 1727, second coach) stopped upright and at a right angle to the track, with the front end near the rear end of the third car. The third coach (No. 1729, fifth car in train) stopped upright with the front end near the rear of the third car; the rear end was derailed to the north and stopped on the track structure of the Westbound. The last derailed car was the fourth coach (No. 1734, sixth car in the train), which stopped in line with the track; only the lead truck was derailed to the north. Considerable damage was inflicted to the first four cars, while the fifth and sixth cars were just "somewhat damaged."

It was in the third car (*Powhatan Arrow* coach 539, which now belongs to the Watauga Valley Railroad Historical Society and Museum) that most of the rescue drama took place. Two recently-discharged soldiers from

the Aberdeen Proving Grounds, Don Yocum and Darrel Hilton – both on their way home to Oklahoma after completing a three-year Army tour of duty – helped passengers escape up the steep incline by using the seats as stair steps. Both were about half-asleep when they heard glass crashing and saw a woman sailing through the air!

Two that were in the car were Mrs. Beatrice Hughes, 32, and her mother, Mrs. Mamie Miller, 52, both of Lynchburg, Va. On their way to visit a sister/daughter in Toledo, Ohio, they were curled up in seats facing each other. "When I awoke, as the train lurched, I had been thrown about middle-ways of the coach, which had gone partially into the river," Mrs. Hughes related. "My mother was also thrown down the aisle.

"We got up and a man told us to get out because the coach was leaning down the bank and might turn over again. Soldiers and a civilian were helping everyone to get out. There appeared to be 20 to 25 people – I'm not sure – in the car." she was quoted in the aforementioned *Huntington Advertiser* story.

Miraculously, only 32 people were hurt, none critically. Fifteen ambulances answered the plea for help. Most of the injured were taken to Williamson Memorial Hospital or treated at local physician's offices.

Convenient to the gruesome scene was the Cedar Branch, which bridged the Tug River and extended 2-plus miles up to Majestic Collieries, over in Kentucky. Many of the injured –and including the lifeless body of engineer Willard –were loaded onto a motor car with three trailers and taken to the Kentucky side, where a road existed to expedite transport.

Officials of the N&W and a special relief train that was brought from Williamson soon arrived, carrying medical personnel and scurrying railroaders who answered the mournful disaster signal that issued forth from the N&W roundhouse whistle. About 40 rode this train back to the Mingo County seat for shelter, while the five cars on the rear (Tavern No. 1722, Diner No. 1022 and Pullmans, *Scioto County, Mingo County* and *Ohio State University*, in that order) that remained on the track were pulled back to Bluefield with about the same number aboard.

Rail traffic was rerouted over the Chesapeake & Ohio via Kenova-Huntington-Barboursville-Logan-Gilbert-Wharncliffe until a main line could be restored 12 hours later. Replacing only a kinked, 39-foot section of rail and three crossties and performing some alignment work was all that was necessary to put the Westbound back in service.

Retrieving the errant No. 611 was another story. After salvaging the other damaged cars, cables on Wreck Car derricks (including diesel conversion No. 514900, built by Industrial Brownhoist in 1923 and now on display at the Mad River and NKP Museum in Bellevue, Ohio) snapped like thread when efforts were first made to pull the engine to the top of the steep embankment. Large wooden blocks had been placed under outriggers extended on the south side to stabilize the cranes and keep them from upending during this particularly dangerous time, when a carman could have easily had his head lopped off or been cut in two!

A bulldozer was then put in operation to cut a roadway up the bank and alongside the locomotive. Next, panel track was constructed in hopes of winching the engine and tender back up to the roadbed. Finally – more than four days after the mishap, the task was completed about 1:30 on Thursday afternoon of January 26 and No. 611 was towed to Williamson.

Several interesting sidebars beg to be told about No. 611's Cedar wreck:

- It was the last major mishap of a steam-powered passenger train in America. On the national scene, it was somewhat overshadowed by the January 22 wreck of Santa Fe's *San Diegan*, which took a 35-m.p.h. curve at 70, killing 29.
- We can thank veteran photographers Tom Wellman of the *Williamson Daily News* and Maurice Kaplan of the Huntington Publishing Co. (both of whom, co-author Hensley knew) for what few images we have. Indirectly nearly becoming the second and third fatalities, Kaplan and his pilot, Carlton Clark, almost crashed in attempting to get the aerial, when they took off in a snowstorm and narrowly missed a tree on a ridge! "I don't know how close we came," Kaplan said, "but I think we got some leaves on the wings!" Like the lawman who always gets his man, the determined lensman took off again the next day and flew 65 air miles to get his famous picture.
- One passenger, John Adams, a 28-year-

old student of Southern Baptist Theological Seminary, claimed to have had a premonition about the wreck in a dream. Screaming as he woke up in a jostled Pullman berth, he held baby Carolyn Ruth, while his wife, a registered nurse, tended to injured passengers. The next day, they boarded a Trailways bus to Huntington and continued their journey to Louisville from Tri-State Airport via Eastern Air Lines!

- Talk about devotion to duty! Until he was relieved, Charles F. Puckett, 52, of Ashville, Ohio, who was supervisor of mails on the train, stood guard over his charge for nearly 12 hours with a broken hand.
- As if the N&W didn't have troubles enough, a tank car carrying 8,142 gallons of gasoline exploded at the east end of Williamson Yard on the evening of January 24, scorching a box car and two empty coal hoppers in the process. General Yardmaster John E. Phillips called the damage "negligible."
- And, not surprisingly, there always has to be a hero involved. Here, it was Assistant Trainmaster Dick Dunlap who later became Norfolk and Western's last president and was dispatched to the scene from Kenova. Legend has it that he dove into Tug River and placed the cables underneath and around No. 611. No fan of steam, he always deadpanned that this was an action that he regretted for the rest of his career...

Officially, the reason for the Cedar wreck was cited by ICC Accident Report No. 3676 as "excessive speed on a curve." Under today's governing Federal Railroad Administration, the cause code likely would have been placed under "human error."

Williamson Daily News

"IN THE HEART OF THE BILLION DOLLAR COAL FIELD"

Volume 44—Number 19 | United Press | Williamson, West Virginia, Monday Afternoon, January 23, 1956 | TWELVE PAGES | Price Five Cents

N&W Passenger Train No.3 Wrecks At Cedar; Engineer Is Killed; 25 Others Are Injured

Engineer Says He "Blacked Out"

29 Dead, 142 Hurt In Wreck At Los Angeles

Many Treated At Wreck Scene

Engine Plunges Into Tug River; 5 Cars Follow

The Norfolk and Western's passenger train No. 3, "The Cavalier," derailed rounding a curve at Cedar, W. Va., about 35 miles west of here, at 12:50 A. M. today, killing its engineer and injuring upwards of 25 crewmen and passengers, including the fireman.

The engine plunged down a 40-foot embankment into Tug River, carrying with it three express and mail cars and two passenger coaches. Six of the other passenger coaches remained on the rails, but all passengers aboard the westbound Norfolk to Cincinnati train were jolted by the terrific impact of the crash.

Hospital authorities in Williamson today said that nine persons were hospitalized and 12 received outpatients' treatment and released. Another Williamson couple received treatment at the homes by a private physician while still another patient was given treatment at the Stevens Clinic at Welch ... a section of the train

9 Hospitalized, 14 Treated By Local Doctors

Six men and three women, some of them ...

Initially, it was speculated that engineer Willard, who hired in 1909, might have suffered a heart attack before negotiating the stiff curve. An autopsy was conducted by Dr. W. W. Scott, assisted by County Coroner Tabor Ball who doubled as a local funeral director. A Columbus, Ohio pathologist was also called in, but the results of the tests were inconclusive. Willard likely was scalded to death by the steam or suffocated when his lungs were seared by vapor escaping from a broken pipe.

Perhaps only Ernest Hoback, who passed away in 2000 – and whose hair turned a premature grey almost overnight after the wreck – holds the key to what actually happened with engineer Willard: "We registered the last signal (*Clear!*) with one another and I don't remember much after that until I 'came to' in the river. He just had something on his mind other than railroadin' that morning."

After heavy repairs at Roanoke Shops, No. 611 was returned to service from the wreck in June 1956. On November 18 of that year, while running lite from the engine terminal to Portsmouth Passenger Station in what well may have been her last crossing incident, she was slightly damaged after striking a 1952 Buick at Lincoln Street.

By now, it seemed that No. 611, indeed, was leading a charmed existence. Maybe some higher power was watching over her. Rest assured that many railfans have offered prayers upon her behalf!

Even before the Js were removed from regular service, 611 — like all of her sleek sisters — received an ungainly brakeman's cupola. The application was made at Shaffers Crossing on February 11, 1958. In addition, all the Js

By this time, May 1958, spit and polish was no longer necessary in the minds of management. So 611 appears to have not been scrubbed down prior to going back out with Train 25, the now downgraded *Powhatan Arrow*. The fireman has the steam up and ready to head west from Norfolk Terminal Station. There are three extra coaches up front this day, perhaps handling a military move, or moving maintenance of way laborers back home for the weekend. *H. Reid photo*

were equipped with connections to make them compatible with auxiliary water tenders.

After the Christmas mail and passenger rush late that year, she was unceremoniously stored at the head of a line of four Class Js at Bluefield. We believe the others were 606, 608 and 609, as all three of these would be retired at Bluefield barely five months later on May 22, 1959.

There was still some life; however. At some point—no one seems to know exactly— the No. 611 was fired up and moved to Roanoke, perhaps about the same time the others were retired. Once in Roanoke, she went into the roundhouse at Shaffers Crossing and was stored. Likely, a request had come in for an excursion train with a Class J, as there was a trip run to Bluefield and around Dry Fork in July, powered by Class A No. 1240 and Y6b No. 2174. Why 611 did not power the train is unknown, perhaps a larger consist than the Class J was allowed, maybe an out-of-date air brake inspection and the Class A was still serviceable; again, nobody seems to know.

However, sometime in August a request came in for an Appalachian Power Company excursion to Norfolk for their employees. With a hydroelectric dam and lake named after him, an APCo vice-president named Graham Claytor Sr., perhaps, requested a Class J. Now why was that done one might ask? Simple, Appalachian was a long-time, large coal-consuming customer and Claytor just so happened to have two adult railfan sons by the name of Bob and Graham Jr.

The exact date of the trip remains a bit of a question, one photo near Montvale dates it to Sunday, August 30, but no return date is specified. Regardless, it was mostly coaches and perhaps a diner on the train, 611 powered it both ways. History is fuzzy, there are also photos of 611 and train west of Roanoke at Shaffers Crossing and, again, at Pepper, which is dated about the same time. Both show the same train, as well as another image that shows the 611 eastbound, with what surely appears to be the same consist because of the aging paint on some cars. These raise the question whether there was a train operated to Bluefield for Appalachian Power as well.

Regardless, 611 went back to slumber at Shaffers Crossing again after the brief flurry of activity. Next up was another fan trip, this time on the east end of the railroad. Meeting an excursion from Washington D.C. that came via Richmond and Petersburg, the 611 powered the train on to Norfolk on Saturday October 18, 1959. There was one spectacular photo run, with lots of black smoke and high-speed running that delighted the passengers, but most knew that it was their last time with a Class J. The 611 returned over Blue Ridge with perhaps a rider coach, but otherwise light, and crossed the famous grade in the darkness.

The following weekend was a contrast of weather, cloudy, rainy and suitable for a funeral. Saturday, October 24, was the grand farewell: 611 to Bluefield, Iaeger and Williamson, where the last of the dinosaurs were running out their miles. The train had stopped at Blue-

Both views on this page were taken during the October 18, 1959 excursion from Petersburg to Norfolk and return with 611. A high speed photo runby with black smoke (page 51) was made to the delight of the sold out train.

For many, their last time with a Class J was from Petersburg to Norfolk and return on October 18, 1959. Our photographer, who rode the train has hurried up front to capture some photos as the 611 backs onto her train. That afternoon, she would be headed into the setting sun, uncouple, then head to Crewe, and finally Roanoke.

field on its way west to take water, then on to Williamson for servicing. The fans explored the big terminal, looking at stored steam— some barely seven years old and already awaiting its date with the scrapper's torch.

It was a wake on the ride back. A stop was made again at Bluefield for water, this time at the east end of the station. The water plug did not want to cooperate; months of disuse caused it to be cranky. But, the tank was finally filled in the darkness and the last ride commenced, the steamboat whistle crying in the night for the many crossings and towns. Finally, down through "Salem Level," Assistant Road Foreman P. C. Cabiness (with engineer Noah Blevins firing) let her open up for one last romp and really put on the show in the dark at 60 m.p.h. The train quietly snaked through the yards and pulled into the station for one last stop.

As the passengers left, the 611 backed up through Track 4, headed for Shaffers Crossing and an unknown future. Officially, a notation in the N&W's Record Of Equipment documents that she was retired three days later, on October 27, 1959.

••••

1959 Memories of Steam, Steel and Snow

Aubrey Wiley

Each of us has some dates that we will always recall. We will remember where we were and what we were doing when we learned of President Kennedy's death; the *Challenger* disaster; and that awful Tuesday morning, September 11. But the fall always brings back very fond, thrilling and good memories for me. Every October 24th, I remember that date in 1959, where I was and what I did. It was a Saturday.

I was 16 then and well before dawn, my dad drove me to the Lynchburg, Va. bus terminal to catch a bus to Roanoke. It was a promising day ahead of me. I had a new camera loaded with a roll of black and white film. I even had a spare roll. I had a leather shoulder bag, a camera bag. That little camera was lost in the immensity of the bag, among my sandwiches and thermos of hot chocolate.

My mom had cautioned me many times about breaking the thermos she packed for me. Remember how easily they would shatter? I could handle it, I was going to ride a steam excursion.

The Roanoke bus station was a mere couple of blocks from the Norfolk and Western station in Roanoke, but for even these then-skinny legs, it seemed terribly far that morning. As I remember, we were to leave at 7 a.m. with our destination being Williamson, W.Va. The morning was clear but cold. I first walked to the front of the train and checked out the engine but it was still too dark for my camera.

Other passengers with nicer, more adjustable cameras were up front recording the beauty of our engine, the 611. It was to be her last run that day. Harry Borsum's "Rail Museum Safaris" operated the excursion. I had ridden on an earlier trip of Harry's on July 11 of that same year and I knew it would be a first class train ride. Our train was 12-cars long: There were two open gondolas at the

Resting at Shaffers Crossing roundhouse, a very dusty and streaked 611 is looking pretty forlorn. Is her career over? It might have seemed that way this hot July day, as the Class A No. 1240 is ready to power an excursion train to Bluefield and back. The 611 would await its first farewell for another day and trip. *Aubrey Wiley photo*

On October 24, 1959, the 611 looks almost like old times. Top left: She cuts off her train at Bluefield and moves smartly west before taking water. Bottom right: A sight at Shaffers Crossing that morning was something a lot of the folks there would never have visualized, the 611 under steam for one last time. Bottom left: At Bluefield ready to head west to Iaeger and Williamson for one last time. *Aubrey Wiley photos*

rear with seats running the length and (having) gangplanks connecting them; In the consist, according to my notes, were a combine car for recording artists, a tavern-lounge and eight coaches, some old, veteran heavyweights.

I had dressed warmly, knowing that the train's speed and my determination to ride outside in a gondola would make it even colder. As the train departed on time, I was in the rear open car. No vestibule Dutch door for me; I had to have the freedom to drift from one side to the other as the train rounded curves! I remember my thoughts from my departure time. Little did I know that the same logic would be utilized decades later by the company.

Our train strode out of the station like a show horse making its best performance. I felt privileged to be on board for such a special run. In the dim dawn light, I waved to all who were watching us at each crossing in Roanoke. It wasn't long before the sun was high enough for pictures and with 611 having such a long train, I could see the front end often as we chuffed westward.

I spent the entire trip hanging over the sides of a gondola, soaking in every site, sound and smell. I knew this was my last opportunity to have these experiences, so I would not be denying myself. I got cold — very cold — and I knew I might become ill. That was OK, because I knew I would recover, but I would

Backing down past her train, the 611 is in search of a working water plug at Bluefield; without active steam for some months, the facilities were harder and harder to find. *Aubrey Wiley Photo*

never again have such an opportunity. To this day, many years later, I can close my eyes and experience those impressions still. The day became colder and the steam was ever so pure and white. When the train reached an uphill grade, the sound of the exhaust was sharp and crisp like no other sound. And the smells, they were indescribable. I wore goggles and I had a scarf tied around my neck. My hat was a striped N & W cap, of course. Occasionally, cinders would make their way through all of my armament and into my eyes.

As we ran out our day's future, we often met freight and passenger trains, all powered by GP-9 diesels. They would salute as we passed. The sky had become overcast as we stopped at Bluefield. We were allowed pretty much free run of the place, except for tracks holding outbound trains. We were all warned that when it was time to leave, the train would leave! I walked over to the roundhouse area and turned to watch our diner being switched out and replaced.

Soon after departing Bluefield, I overheard someone nearby speaking with a New Jersey accent, that we were entering the "N & W subway system!" Actually, most everyone on that train had some sort of Northern accent, I noticed. The train ran slow as we twisted through one sharp curve after another – a constant sequence of tunnels, curves, bridges and more tunnels.

The train stopped at Iaeger for us to stroll about and make pictures. It was still overcast but I tried to get some pictures of the few live steam engines there, all Y class. As we made our way to Williamson, it started to snow. What a wonderful coating it made in my memory. The brilliant fall colors, sugar snow on the ground and the right-of-way, a colorful train and that beautiful streamlined steamer laying pure white smoke along the top of the train! As it slowed for the sharper curves, the sounds of the working engine were reduced and the shrill squeal of the wheel flanges overwhelmed all other sounds.

Williamson held a solemn site for me; lines and lines of repetition. Dozens of N & W's finest steam engines were dead and soon would feel the heat of a scrapper's torch. There were S1as and Ys. From a distance each locomotive had lost its identity in the rhythm of dark shapes. They were just black mounds sitting atop wheels. I stayed in the open gondola this time. I saw nothing within that dreadful site to beckon me. I do recall being impressed by a huge overhead bucket system I could see in the distance. The coal was mined on one side of the river and carried to the other for processing.

As we pulled out of Williamson, I became aware of the dread ahead. The 611 was heading to its last port of call. Soon gray skies gave way to darkness. We were several hours behind our schedule. How foreboding! Nonetheless, I continued my vigil from the gondola, save going inside long enough to get a Coke to wash down my last peanut butter and banana sandwich. On the bench behind me, two men were discussing other 4-8-4s they had witnessed on other railroads. One had ridden a recent "Iron Horse Ramble" behind a Reading T-1. The other talked of how he wished he had seen a C&O J-3a in action. Then they agreed that any road's 4-8-4 would beat New York Central's 4-8-4 in appearance. I thought to myself, "I wish I knew more about what they were talking about." I knew little beyond the roads with which I had contact.

From Bluefield east, the trip was uneventful. I still hung over the side of the gondola as far as possible in the cold wind that night, soaking in the thrill of seeing the glow from the engine as it rounded the curves on my side.

When we eased to a stop at the Roanoke station, I didn't want to rush forward to see the engine. I thought it would be almost like going to the bedside of a dying friend. I wanted to remember the 611 as the beautiful engine it was, the engine that had been so alive and strong all day.

My parents met me on the platform and had so many questions I couldn't answer. As my dad drove us home, I still couldn't give sensible answers. I leaned against the door of the back seat of our three-year-old DeSoto and slept a good sleep. It was October 25th now and I hadn't broken the thermos.

EDITING NOTE: According to an article in the November 1959 issue of the *Norfolk and Western Magazine*, 450 rail enthusiasts from 25 states rode that October trip. It mentioned nothing about being the 611's last run.

On the outbound trip, a stop was made at Auville Yard in Iaeger. "There were whoops of joy from the loaded gondolas on the rear as the train backed slowly up the Dry Fork Branch and a dozen or more Y-6 engines came into view, at least eight of them under steam," the story quoted. "The fans spent half an hour there, taking pictures from every angle and some new ones."

Another who was on the trip was Larry K. Fellure. then of Huntington, W.Va., now a retired CSX Dispatcher and part-time Disneyworld steam engineer. While at Iaeger, he recalled that railfans were given free reign to climb into the Y-6 cabs and tug on unattended "hooter" cords. "'The Great Whistle Storm' on O.Winston Link's *'Thunder On Blue Ridge'* recording paled in comparison to it!," Fellure said.

Continuing with the report, the N&W publication stated about the "going" trip : "Railfans, men and women of all ages with children of every size and from every economic level are a hardy breed. They crowded the open cars for the thrill of riding through Elkhorn Tunnel and when the train rushed from the west portal into a driving rain, there was only good-natured laughter as they lined up single file for the shelter of the last coach." The "subway system" voice was possibly that of O. Winston Link.

••••

611 is ready to fill her tender at Williamson. The bucket system is very visible here. Note the canteens stored on the adjacent track. *Bob's Photos Collection*

The End of Steam, But Not Quite

Just who saved the 611 is a matter of conjecture. Many have tried to claim credit – O. Winston Link among them, who professed that he lobbied former President R.H. Smith to keep two N&W locomotives of each class in a no-longer-needed roundhouse, similar to what PRR did at Northumberland, Pa.

More likely, her primary protector was the elder of two brothers who were about to make their mark on the American railroad scene. Less than a week before 611's last run, Graham Claytor Jr., then with the Washington law firm of Covington & Burling, wrote Stuart Saunders (an old acquaintance and fellow Harvard Law School graduate) – in a four-page, October 19, 1959 letter – with hopes of extracting a promise from N&W's new boss that the locomotive would not be scrapped immediately.

This where the effort to save 611 all started. 611 has arrived back in Roanoke with the Appalachian Power special train. Back in the cars, Graham Claytor has spoken to Stuart Saunders and would follow up with his first letter the next day. *Norfolk and Western photo*

After congratulating Saunders on some recent corporate milestones (ICC approval of the N&W-Virginian merger), Claytor said: "In a final effort to accomplish a considerably lesser feat, but one that also has been said 'couldn't be done,' this is to obtain at least a stay of execution on your last Class J locomotive No. 611." Claytor had started this letter campaign with Saunders about 611 within a day of the Appalachian Power trip in August.

Graham Claytor eloquently argued that the out of pocket costs of storing versus scrapping the locomotive were entirely "separate" issues. He then went on to quote ICC regulations regarding the rostering and inspection of stored locomotives, taxation and depreciation issues and even pointed out that the Louisville & Nashville had to lease a Illinois Central steam engine to commemorate an important Centennial. He pointed out that No. 611 was already in excellent operating condition, that a "spare" water tender could offset long-run excursion logistics and that an already-on-the-property supply of component parts could keep it going.

Considering the merits of storing instead of destroying the engine, he continued: "A decision to cut it up will necessarily be irrevocable; on the other hand, a decision to put it aside in its present operating condition can at any time be reversed and it's scrap value realized, if it turns out that a continuing storage is no longer desirable or becomes too expensive. Accordingly, not even an implied commitment is sought that it will be financially desirable or feasible to actually operate the engine on future fan or excursion trips."

Barely a year before, a Class J on a passenger train wouldn't have been a special occasion. But by late August 1959, an Appalachian Power special would bring noted photographer H. Reid out for a late afternoon view of last-J standing, No. 611, making good track speed at Poe, Va.

While riding in the cab of the big 4-8-4 on her dashing, October 18, Petersburg-to-Norfolk excursion, someone from the N&W Operating Department told Claytor "This locomotive is in first-class mechanical shape and was good for another 40-50,000 miles at least." These points, he also made to Saunders.

"The long-term public relations value of keeping one of these famous locomotives would more than equal the modest storage cost," he continued. "The Class J represented the epitome of steam power. It was a product of Roanoke designers and Roanoke builders. Although there were only a dozen (sic) in existence, in the last few years it has properly become the most famous steam passenger locomotive in the country."

In closing, Claytor wrote, "On behalf of the railroad industries and N&W's many friends, I seriously plead for at least postponement for further consideration of the necessarily, irrevocable decision to destroy this engine." He signed the letter "with hope."

A rumor persists that protective employees at Shaffers Crossing roundhouse draped tarpaulins over her and stacked empty 55-gallon drums around her. Sequestering the locomotive in such a manner would keep her hidden out of sight and out of the mind of Stuart T. Saunders, they reasoned.

Presumably, Saunders was not in a malicious mood at the time, as he had just spearheaded the 1959 merger of the Virginian Ry. and already had his sights set on the acquisition of the Nickle Plate and Wabash, which occurred in 1964. Prior to that, he had moved on to fry bigger fish in Philadelphia a year earlier.

Also, working quietly behind the scenes was the future (1967) Southern Ry. president's younger brother. Robert B. Claytor, a humble man who sang tenor in his church choir, had joined the N&W's legal department as a solicitor in 1951. Bob knew that he would have to keep his visibility low, because there was little sentiment for steam within the company culture and it was not savvy for a rising young officer to be a card-carrying railfan.

But both brothers had been heavily involved with the 1957 National Railway Historical Society Convention that was held in Roanoke. And, often in the evenings before the end of steam, Bob would slip out to Boaz and ride the Blue Ridge Y6 pushers with his buddy, Lake Walrond; this paid off handsomely for the personable Norfolk Division engineer, who also operated an automobile repair business out on Route 460 east of town.

On November 3, 1959, Saunders responded to Graham. "A decision has been made to retire, but not scrap, the engine at this time," he wrote. This had to be considered as, at least, a moral victory.

The next year, Graham sent a note on April 26 to brother Bob that a rumor was circulating that the 611 was about be scrapped. He urged him to "check on 'our engine.'"

During that summer of 1960, heads of the C.P. Huntington, Old Dominion and Tidewater chapters of the NRHS were aggravating Saunders with requests to run excursions. Graham Jr. put out a memo to them that they needed to cease and desist while he pondered the locomotive's ultimate fate. Underscoring this was a note that C.E. Pond wrote to Bob Claytor on June 12, 1961, that it was "best not to write Saunders or Wyatt," the latter being Harry C., who was N&W general manager at the time.

Graham Clayton was already hard at work on convincing Stuart Saunders to save the 611 when he rode the cab on the October 18, 1959 excursion,

In a December 2 note to Bob, Graham wrote that he was informed that the railroad had turned down a $5,000 offer for 611 and the unidentified, prospective buyer (Could this have been Winston Link?) was told the locomotive was "not for sale." In addition, he had personally inspected her during Thanksgiving, finding the locomotive in generally good condition and adding, "I'm hopeful that, when the appropriate time comes, we will see her on the road again!"

Bob's underlying fear was that an effort would be made to make the 611 forever inoperable by taking a torch to the firebox. On orders from top management, this had been done to several retired locomotives. In 1962, Robert Claytor sent a personal note to C.E. Pond requesting that he be made aware of any such decision before the actual work was to be done.

As proud as N&W had been of its steam power and unlike Pocahontas Road competitor, Chesapeake & Ohio, it made no conscientious, organized effort to preserve its products. We are, indeed, lucky that one each of the three examples of the "Modern Coal Burning Steam Locomotive" inadvertently survived, considering that such important company points as Norfolk, Petersburg, Crewe, Lynchburg, Shenandoah, Radford, Pulaski, Bristol, Norton, Iaeger, Williamson, Kenova and Portsmouth never received a displayable engine!

The demise of steam (with its resultant lay-off of shop and roundhouse forces), the closing of the American Viscose plant and an economic recession were hurting the Roanoke area. Things were quiet for a time, the many retired steam locomotives were feeding the furnaces of the steel industry, while the "New N&W," touting itself as "The Nation's 'Goingest' Railroad," focused on the future with America's most modern diesel fleet.

A scare was experienced on March 4 of 1962, when "Cinder J." (Jack) Stith, president of the NRHS Old Dominion chapter in Richmond visited Shaffers Crossing and found the 611 missing. He was told by local employees that she had been "gone two or three weeks."

Naturally, panic ensued. But instead of being scrapped, she turned up – with some serviceable flue time still left – at the East End Shops on standby service, while one of two big stationary boilers was down for maintenance.

At one point, an estimate was apparently made by the N&W to put the 611 back in service. But at $16,000, it was deemed too much.

Beginning in early 1962, some civic-minded parties had been discussing a bold new plan, that of a transportation museum in Roanoke. On April 2, 1962, Roanoke City Council passed a resolution to undertake the completion of this museum, to be built in an underused city park in the Wasena section. The idea, which was the culmination of a dream of City Manager Arthur S. Owens, was catching on quickly.

When news came of this, Graham could see where the 611 would fit in. He was in correspondence with City of Roanoke officials, pitching having the 611 as a museum piece, but operable to haul mainline excursions as well.

In late May 1962, Stuart Saunders abruptly decided to give the 611 to the "Star City" for its fledging museum. In a memo of the 28th to all involved, the elder Claytor wrote: "Suddenly and without notice, the Norfolk and Western has just announced that it is giving the 611 to the City of Roanoke. It is not interested in restoring the locomotive to operating shape. The decision to make the transfer has been made public and is therefore, irrevocable."

Even then, Graham continued his campaign with the municipal authorities to see the locomotive under live steam. But, there was no one with an equal vision to help raise the necessary money. He did donate $500 (approximately $3,900 in 2015 dollars!) to the city for her upkeep. As a condition of his gift, he chastised Arthur B. Stone who headed up the committee, not to let the engine rust away into another example of the typical city park steam graveyard.

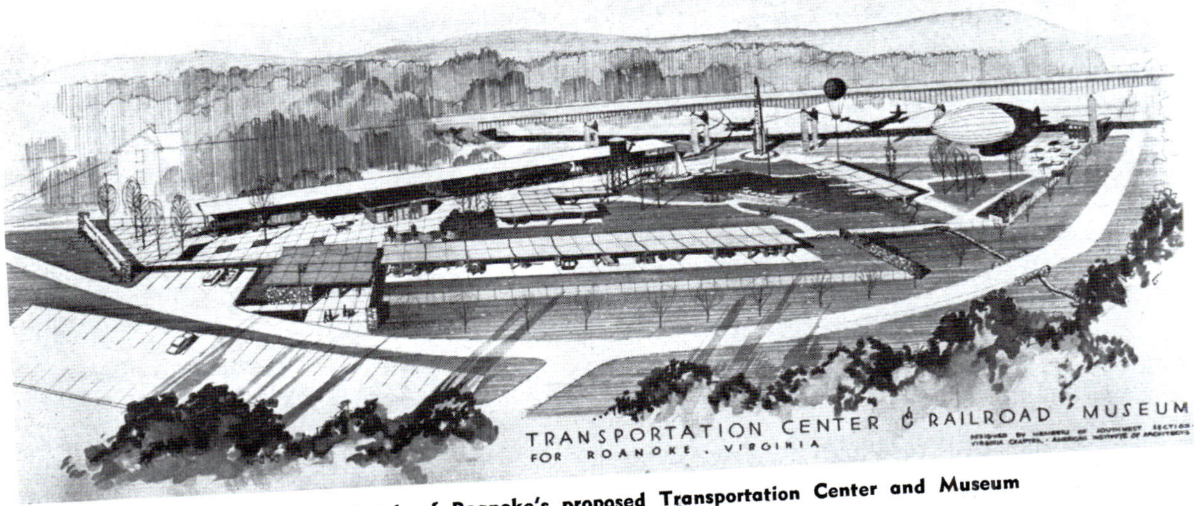

An architect's sketch of Roanoke's proposed Transportation Center and Museum

Roanoke's Transportation Museum

A spectacular Transportation Center and Museum at Wasena Park is developing in Roanoke.

The center, to feature two Norfolk and Western steam locomotives in addition to a variety of other transportation equipment, is expected to be opened to visitors on May 30. The opening is scheduled to coincide with the visit to the Norfolk and Western and Roanoke of the famous Civil War locomotive, *The General*, which will make an exhibition run over our railroad under its own power as announced on page 155.

The Transportation Center is the result of the long and persevering work of an energetic committee headed by Arthur B. Stone, retired N&W chief engineer. This group, which was organized in January, 1962, has been at work actively on the project since that date.

Besides the steam locomotives, the museum will provide a permanent exhibit area for a caboose, trolley car, a dirigible, airplane, balloon and many other units exemplifying transportation history and progress. The exhibit area also will include a small lake with water craft, an actual railroad station, water tank and a rocket missile.

James Trout, draftsman in the Engineering Department, is chairman of a finance committee to raise funds for the transportation center. Wiley N. Jackson is co-chairman.

Members of the Transportation Center committee, in addition to Messrs. Stone, Trout and Jackson, are: Arthur S. Owens, John L. Wentworth, F. Cletus Broyles, Rex T. Mitchell, Jr., Werner K. Sensbach, C. E. Pond, William A. Lashley, Frank B. Mundy, H. A. Fisher, W. B. Adams, John Chappelear, and Frank Hill.

The April 1963 *Norfolk and Western Magazine* had this article on the soon to be Transportation Museum.

Top: A long time supporter of the Roanoke Transportation Museum, Graham Claytor had not forgotten Roanoke once he became president of the Southern Railway. He is shown here speaking at the presentation of coach 903, *Pulaski* (from the 1941-built *Tennessean)* in 1967. *Bottom*: Brother Bob was also a supporter of rail preservation. He is shown at Crewe on May 1, 1971 while NKP 759 is being coaled. Assuredly, both would have liked nothing better than 611 running that day. *Both: H. Reid photos*

611 looks pretty nice in this summer 1968 view at the Wasena Park Transportation Museum. Even as a museum piece, she was a popular backdrop for family photos. *Art Bixby photo.*

Then, in the spring of 1963, after the organization was formalized in March, things started happening at Wasena Park and a grand opening was scheduled for Memorial Day. The 611 was rolled into the Paint Shop and again she glistened in fresh black and Tuscan red with gold trim.

In the halcyon days leading up to the opening of the Museum, she was gingerly rolled over the Roanoke Belt Line and parked on a newly-constructed siding built just for her. However, she was not to move under steam.

Steam was, however, at the Museum opening, in the form of the L&N's *General*, the historic, "Great Locomotive Chase" 4-4-0 touring the states as part of the 100th anniversary of the Civil War. Graham was also on the move, as he joined the Southern Ry. Law Department. He still maintained his interest in the Roanoke project, later donating several pieces of equipment to the new museum.

If you haven't figured it out by now, both he and Bob had dreams of the 611 running again. Steam was something that they had in their blood and it was not many years after Graham became senior vice president-Law that the Southern Steam Program, developed under the guidance of entrusted associate Jim Bistline, was born. And, those trains did occasionally visit parts of the N&W.

Starting with Paul Merriman's 4501, an ex-SR 2-8-2 that ended serving the Kentucky & Tennessee short line as it's No. 12, the Southern utilized quite a stable of steam excursion engines before the 1982 merger with N&W that resulted in Norfolk Southern. An excellent history, called *Steam's Camelot* and written by *Trains* magazine editor Jim Wrinn while he was with the *Charlotte Observer,* tells of those formative years and beyond.

One such excursion took place on June 17, 1967, when East Tennessee & Western North Carolina 2-8-0 No. 207 was borrowed in conjunction with a dedication of historic equipment (including Wabash E-unit 1009 that was the 10,000th diesel built by EMD) that took place at what was then known as the Roanoke Transportation Center and Railway Museum.

That's when little Preston Claytor (son of Robert and just shy of seven-years-old) got his first whiff of coal smoke. Father Bob had told brother Graham about the two little Consolidations based out of Elizabethton, Tenn. and he traded two diesels for what would become (with the restoration of their former SR numbers), the 630 and 722 (nee, ET&WNC Nos. 207 and 208, respectively).

As his uncle's Southern program developed, a teenage Preston learned how to cut coal and eventually fire. Another notable trip took place during the second NRHS convention that was held in Roanoke, when leased Texas & Pacific 2-10-4 No. 610 was doubleheaded up Blue Ridge with the 4501 on Saturday, September 3, 1977. On a next-day, Sunday solo sprint up Christiansburg Mountain and on to Bluefield, the big oil-burner's exhaust bounced off the roof and blew back with such a force that it extinguished the fire as she stomped through the Westbound (north) track of the twin Montgomery Tunnels. In what was the most hilarious scenario witnessed in all his years of chasing trains, co-author Hensley will never forget the startled, deer-in-the-headlights-look of the entire, soot-faced engine crew as they faced the firing line of photographers after drifting to a stop upon exiting the billowing bore. It was like a minstrel show in motion! Not until afterward did the seriousness of the potential consequences set in.

Time to Wait, and Wait Some More

However, it was not time yet to bring out the 611. Rumors floated, but none seemed to bear any substance. Time passed for the fans, if slowly. Hope does spring eternal!

As a new decade arrived, the railroad world would be changed forever: the end of regular passenger service and formation of Amtrak was announced. The sad news came that the N&W route would not be part of the new system. That would mean, on May 1, 1971, there would be a last run of N&W's passenger trains.

Obviously, someone (Bob Claytor comes to mind!) suggested that it was right to send a proud tradition off in a grand farewell. We are certain that Bob would have loved nothing better than to have 611 on the head end of that last No. 4 out of Roanoke, but again that was not to be. Instead, Nickel Plate Road 2-8-4 No. 759 had run a couple of trips on the N&W in Fall 1969 and was powerful enough to handle the last run. The 759 was run to Roanoke with freight; the tender was sent through the Paint Shop, where "Norfolk And Western" was stenciled and everything was ready.

759 is blasting out of Roanoke with the last N&W passenger train, No. 4, the *Pocahontas* on May 1, 1971. She had been relettered for the occasion and a Class J whistle was applied for the trip.

That sunny spring day appeared to be anything but a sad occasion, as crowds flooded the Roanoke Passenger Station for one last time. On display was the cosmetically-restored N&W Class A, No. 1218, recently acquired from Steamtown; nobody there could have guessed that the 1218 would one day steam again too!

Instead of having the usual, few scattered riders, the *Pocahontas* was a full train that day. To recreate a bit of the past, Bob Claytor provided his N&W Class J whistle to replace the somewhat anemic NKP quill on the 759 — the first time in almost 12 years that a J whistle was heard in Roanoke.

The 759 hauled the big train east with trackside crowds all along the way. A photo in the Roanoke paper showed a rather uncomfortable-looking N&W President John P. Fishwick sitting on the engineer's seatbox of the big Berkshire. A lawyer by trade and cut from about the same temperament as Stuart Saunders, "Jack" Fishwick was said to have asked as the train passed by hundreds of people at Bedford, Va., "Are all these people here to see us?" The answer was "yes." He put up with the infrequent steam trip on occasion, but would never sanction such a thing as a regular event.

In 1970 rumors were afoot that the 611 was soon to run again and the Roanoke Chapter of the NRHS was even putting together a plan to make it happen. Bob and Graham Claytor were, as expected, supportive and even tried to pull it off as early as the next year. A meeting was arranged between chapter officials and Fishwick, seeking his permission to perform a hydrostatic test on 611 prior to her restoration. His response was, "Gentlemen, you can test anything you want, but I can assure you that it will never run on my railroad." Thus ended the discussion.

After the demise of the N&W passenger service, there would still be occasional steam trips on the N&W. But that 611 hope was still there, just a bit dimmer than it had been in 1970.

Secretly, in January 1981, an incognito N&W crew was at the Transportation Museum, looking over the 611. This was being done as quietly as possible, on a Monday, when the Museum was normally closed. Though there were telltale signs that something was about to happen, neither the Museum nor the N&W would comment.

However, then Museum president David Helmer might have slipped up when asked by co-author Miller if 611 was coming out to run. His response was "Well, where there's smoke, there is usually fire."

N&W crews would visit the Museum each Monday for almost a month, looking things over, oiling and greasing the 611 and apparently making her ready to roll out. Then, work stopped for no apparent reason. Nothing was said, about why, but maybe that fellow Fishwick had gotten wind of such and stopped the perceived "foolishness."

THE EMBERS BEGIN TO GLOW

Rumors of restoration began to float around the railfan community during the spring of that year and were published in many fan-oriented publications. The beginning of October brought the news that John Fishwick was retiring and Bob Claytor was moving up as his replacement. Within a week, crews were hard at work on the 611, as had been anticipated under the younger Claytor's leadership.

The official announcement was made on Tuesday, October 13, 1981 that the 611 would leave its home of the last 18 years on Friday, October 16th, 22 years, to the day, of one of its last excursions. The 611 was to be taken to the Southern Railway's Steam Shops at Birmingham, Ala., to see whether it was feasible to restore it to operating condition.

Virtually everyone knew that once the lady had gone to Birmingham, she would be back in steam again, regardless of cost. Several days of hectic activity preceded the official move, scheduled for 10 a.m. Track work was necessary; other displayed equipment had to be moved out and brakes had to be made functional on the 611. To assist in the upcoming restoration, detailed photos of 611, had been produced by co-author Miller during the summer.

Like a Pied Piper, the 611 leads a crowd out of the museum on October 16, 1981. She is bound up on the new track connecting the Belt Line. She is faded and ragged, but barely 9 months later she'd be a showpiece. *Kenneth L. Miller photo*

October 16th dawned as a beautiful fall day. Just before 10:00 am, N&W GP-38AC 4149 and GP-9 750 shoved a loaded hopper up to the tender of the 611 and coupled on. Everything was cleared out and the units started to pull. The 611 rolled out easily, with no creaks and groans except for the track. The only noise from the 611 was the exhausting of air from the drifting valves, maybe a signal that she was already coming back to life!

One thing that had helped the 611's bearings all those years was Roanoke chapter member L. Jack White, one of the mechanical crew. Faithfully, Jack would (frequently) go over to the Museum and carefully lube the 611, just waiting for the happy day.

The yard crew was being careful with so many people watching; there was no room for mistakes. The only snag was when the 611 got to the switch off the Belt Line; the long wheelbase of the J and freshly laid track did not want to cooperate. As the 611 crept back, she started to climb the rail. After considerable discussion, some grease and well-placed tie plates safely facilitated the rest of the move. Another loaded hopper was placed in front of her and the trip to the junction at the west end of Roanoke Terminal continued. By now, the 611 had attracted a group of slow speed chasers, as she was towed past Shaffers Crossing and through Roanoke Terminal.

Once the 611 was at old Track 4 at the Roanoke Passenger Station, the diesel units were cut off, leaving one hopper on the tender for brakes. The newer means of motive power ran around her, removed the hopper from the nose and left the 611 to pose exactly where she had been over 31 years earlier when first completed!

The company photographers recorded the event and the units came back to shove 611 into the Roanoke Shops, where she would be checked completely and prepared for the trip south. Bob Claytor, seeing his dream of years coming true, had ridden the cab of the locomotive from just outside the museum to the station.

On the following Thursday, October 22, the 611 was towed out to Shaffers Crossing and run back and forth on the old break-in track between there and the west end of the yard, to make sure there would be no bearing trouble for the long haul to Birmingham. No problems were encountered and so on a rainy October 23rd, GP-35 222 coupled to the 611 with three loaded hoppers and a caboose and headed west to Bristol.

The trip was uneventful until the 222 lost its footing and stalled at New River, just west of Radford. The train backed to Radford and added GP-35 No. 1366, which assisted in getting the 611 to Bristol at 4:30 that afternoon. The Southern replaced the hoppers with boxcars and a GP-30, No. 554 and then headed south, finally arriving in Birmingham at 5:30 pm on Sunday, the 25th.

The steam shops crew was out on the road with ex-Nickel Plate 765 at the time, but two shifts would begin to work on 611 after the excursion season was complete. Flues and superheater tubes had already been ordered; now taking 611 apart was next.

Bob Claytor had risen to N&W President in October of 1980 and CEO in October of 1981. Speaking about the 611 in December 1981, he said, "Nothing I have done has given me more pleasure since coming to the railroad as taking the 611 out of it's tomb at the Transportation Museum." He emphasized that no trips had been planned yet, but a schedule would be worked up as completion of the 611's restoration was closer.

Work continued through the spring and early summer of 1982, with a target date of Labor Day weekend for the first official trips. The N&W and Southern officially merged into Norfolk Southern on June 1, 1982, making the system one large railroad. What better way to herald the marriage than with a 611 honeymoon?

Those Embers Become Flames

Sunday, August 15, 1982, qualifies as a big day in 611 history. For the first time since 1959, a Class J would be under steam and in motion again. Shimmering in fresh paint, the 611 stood outside the Birmingham Steam Shop at Southern's Irondale, Ala. yard. The steam team posed in front of the freshly-refurbished locomotive and then slowly, she was backed off the shop track, then up through the yards. Behind her tender was an ex-L&N tank from a Class M-1 "Big Emma" 2-8-4, former N&W Mail Storage car No. 1407 and a Southern bay-window caboose.

Shop Foreman Bill Purdie and his shop crew were treating the 611 carefully, stopping every few miles to check the bearings. Their destination was Chattanooga, some 143 miles away. With many stops to check things over, it took three hours to cover the first 40 miles. Without problems showing up, it was time to open her up a bit, easily reaching the 50-m.p.h. limit the crew had imposed. She made quite a sight running through the Northern Alabama countryside. Only minor lubrication problems had cropped up on the trip and 611 would pull a freight train back to Birmingham the next day.

Her boiler a ghostly grey, the 611 is sharing Birmingham shop space with ex-C&O Kanawha No. 2716. The 611's bolier is being sandblasted (outside) and has been rolled back into the shop for the evening. The 611's sand dome, still dented from the January 1956 wreck, is on a stand between the two locomotives. March 1982.
Kenneth L. Miller photo

Veteran Norfolk Division Road Foreman of Engines Frank Collins, an N&W steam-era engineer, was in the cab for the test run. Collins was originally off the Radford Division and a trusted Claytor associate. He was assigned to work with the steam program and stayed until his retirement at the end of 1987.

Homeward Bound

The trip home commenced on Friday morning, August 20th, with the first leg to Atlanta. The next jaunt would be to Salisbury, N.C. on the 21st and finally back home on N&W rails at Lynchburg on the 22nd.

A private, special train with the Claytor family and guests had left Roanoke on the afternoon of the 19th and arrived in Birmingham before dawn on the 20th. The passengers, several jolted awake with the stop at the Birmingham depot, looked out the windows as the special train was backed to the site of the former Terminal Station.

In the dark, the 611 was eagerly waiting. As day broke, she was now in place on the head-end, and impatient to return to her birthplace. Many guests and fans were milling about on the platform. Then, Bob Claytor came out of the business car and walked up front to see his "baby" under steam for the first time since 1959. Tears seemed to well up in his eyes as the early morning sunlight glinted off the shining bullet nose. The chromed handrails sparkled with highly-polished brilliance. He climbed to the cab with a big smile on his face. One could

Test run day, August 15, 1982. The "steam team" poses in front of their newest baby. *Kenneth L. Miller photo*

see it was his dream come true, after many false starts and more than 20 years.

The cab was slightly different than it had been in 1959; the original 8ET brake had been replaced with a modern 26RL stand that allows easier repair and replacement parts. A radio was now in place and the cab interior had been sheathed with varnished hardwood. The chrome-plated throttle and other handles replicated that N&W practice on passenger locomotives, but the N&W had never sheathed the cab in wood. Cab window frames were painted Tuscan red, another change from the original N&W practice of orange window frames, but these were minor things. One could not help but be struck by the amazing beauty of the machine in the glow of that early morning light.

Soon, it was time to depart, the true test of the J, again on the front of a passenger train. Running easily during the day, the 611 was beginning to stretch out and become familiar for the crews. Arrival in Atlanta was in the late afternoon and 611 and company bedded down for the night.

Heading north the following day, everything was becoming routine. A discussion of doing a runby quickly developed into a stop and passengers unloaded to allow a photo session. Bob and Graham Claytor were out as well, video cameras in hand, as the 611 backed up for a good distance. Then, with a jubilant whistle call, a *real* runby at 60 m.p.h. was made. It was not a slow move for benefit of the long line of people, but a real Class J runby, at speed for the 20 or so lucky riders on board.

After tying up for the night in Salisbury, N.C, dinner was served in the N&W dining car. The arrival of Claudine and Elbert Miller from Roanoke was a surprise. They had a Roanoke newspaper in hand, the wire service had sent out a photo of the 611 arriving at Atlanta the previous day. Bob Claytor was excited to see that Roanoke City was proud of her own and invited the couple to join them for dessert.

The next morning, Sunday, August 22nd, was the last leg of the trip home. The train left an hour early as the NS chairman wanted to arrive in Roanoke on time and risk no delays. Running north in the early morning fog and haze, it was a step back into the past again. As riders lingered over coffee or tea in the N&W diner, they watched wisps of steam roll past the window as the J's side rods pumped onward.

At Lynchburg, the train rolled onto N&W track a little after 1 p.m. and proceeded to the metal N&W station at Woodall Rd. near Kinney Yard, which would play a sad part in the 611's history 12 years hence. The 611 and train pulled in and tied down. Those on board wanted her to look her best as she bounded over the Blue Ridge for home. So the water hoses and brushes were brought out and she was properly groomed. Another train, diesel-powered, was on the way east from Roanoke with five cars of dignitaries and invited

"Pappy" Houseman is checking the heat of the bearings after the trip from Birmingham to Atlanta, August 20, 1982. *Kenneth L. Miller photo*

Friday, August 20, 1982, the sun is not even out as the 611 and train awaits departure at the former site of Birmingham Terminal Station. *Kenneth L. Miller photo*

Top left: A water stop at Anniston, Ala. on August 20, 1982 finds 611 with admirers all around. That is future Manager Steam Operation Carl Jensen in the blue shirt speaking with Engineer Frank Collins, wearing his ever present khakis. Just for the sake of contrast, it was hard to believe that it was less than 10 months before, that the 611 was being towed to East End Shops *(lower left)* and just five months before, her cab and nose rested in Birmingham Shops. By the way, on the lower left photo, it is not a mistake: The stripe on the tender IS very crooked; on repainting in Birmingham, it was corrected.
All: Kenneth L. Miller photos

guests, who would stirringly ride back to Roanoke behind the J.

The additional cars were cut in at the rear of the train and a solid red consist headed home to Roanoke. N&W's trademarked "Precision Transportation" slogan was the hallmark of the day, with scheduled arrival at 4:00 p.m. A stop was made at Bedford to kill a bit of clock; exact, on-time performance was the only acceptable answer to her long-awaited, grand entry. The 611 looked serenely at home passing under the N&W's position light signals, all lined green for the CEO into headquarters. Climbing the Blue Ridge grade at a lazy pace, she steamed with no real effort.

A large, excited crowd had gathered at the Roanoke Passenger Station to meet the newest star of the Norfolk Southern steam program, as she rolled into Roanoke exactly on the advertised, Bob Claytor made a brief speech, mentioning again that it was the proudest moment of his railroad career.

Stepping out of the cab onto a specially-built platform, he looked out over the throng and appreciatively quipped, "If we had as many people as this coming to the station in the days we ran passenger trains, we'd still be running them!"

On N&W rails at Lynchburg, the 611 is being washed for the final leg of her trip home. The man in the foreground with the beard is Tim Hickman of Image & Sound Productions, filming the NS documentary *"Going Home"*

"I'm going to introduce to you an old citizen of Roanoke," he beamed. He forcefully stated, "Roanoke born, Roanoke bred and Roanoke proud! Ladies and gentlemen: I give you N&W's Class J No. 611!"

He turned and presented the "finest steam passenger locomotive ever built" to her hometown. The 611 answered with a long, triumphant blast from her steamboat whistle. It was a very intense, emotional moment for two long-time partners: Roanoke and Norfolk Southern.

Excursion Service

We now turn our focus to the 611's excursion season which, by the yearly calendar, lasted longer than her original service.

Once home, 611 was only to rest for a short time before heading off into her new life. The first public excursion scheduled was from Roanoke to Norfolk and return over Labor Day weekend, 1982. The train had been sold out for weeks and anticipation was high. The 611 was rolled out and – for the third time in her life – posed at the passenger station, again for company photographers (and a few other invited guests) on Thursday, September 2. The big day, Saturday, September 4, 1982, was at hand, with a train load of fans and 16 cars.

Rolling out of Roanoke right on the advertised, the train headed east amid many chasers, curious onlookers and many employees or retirees. 611 worked a little climbing Blue Ridge, but it was a larger train than most of the Js handled in regular service.

Graham Claytor, perhaps thinking back to another warm September day some 23 years before, is waiting for a highball at Lynchburg, Va. on Sunday September 6, 1982.
Kenneth L. Miller photo

Things went well and the first photo stop was made at a little community known as Lowry, almost 34 miles from Roanoke. Nearly everyone on board climbed down to have a look at the beautiful J and train. The 611 backed west and with a dramatic pause, rushed eastbound at speed for the passengers. When the train had cleared the crowd and stopped, a spontaneous ovation was given by all alongside the track. The planners knew that they had done something right!

The 611 and train headed on east, with many onlookers at every point along the way. It was reminiscent of the last *Pocahontas* of eleven years earlier, except this was no final run. Making good time, the train arrived in Norfolk without incident and with a trainload of happy passengers. The train was turned on the large balloon track around the yard, while the 611 was serviced and sent to turn on the Lamberts Point turntable, still in place since the end of steam.

It was about two hours later when one of the steam crew came into the hotel restaurant in dirty work coveralls. Bad news – 611 was on the ground. Apparently, the lead track to the turntable had been modified since the J's had regularly used it; now a set of drivers was off and an old-fashioned, plank-board and blocking rerailing was the plan. Understandably, the Norfolk Terminal folks were concerned, after all this was the CEO's train; he was on board and nobody wanted to take a chance of further derailing his unique locomotive.

By the next morning, 611 had been retracked with no consequences, but was still pointed in the wrong direction. Norfolk Division folks were understandably concerned and were afraid of a repeat of the night before. The decision was made to go west with diesels to keep the schedule. The 611 would follow, running backwards to the wye track at Lynchburg, almost 200 miles away. The backup move was restricted to 15 mph, then there were 50 miles back to Roanoke. The 611 and crew arrived about 6:00 am, Labor Day morning, just in time for turning, servicing and returning to the station to go out with a 9:00 a.m. departure to Alexandria, Va.

Reminiscent of regular service, the fast turnaround time was successfully done. At Shaffers Crossing, a number of the facilities like the turntable, roundhouse and lubritoriums were still standing, although the J servicing materials had long since disappeared.

The trip north to Alexandria was without incident and gave the 611 the chance to stretch her legs on Southern's mainline with a 79 mph speed limit. A week later, she ran for filming of a *National Geographic* special. With number boards reversed, she was captured smashing through a mirror for effect. She also handled part of a legislative special, bound for White Sulphur Springs, W. Va., before returning to Roanoke.

About this time the locomotive added to her folklore, when the *"Ballad of the 611"* was released. The 45-r.p.m. platter was written, composed and recorded by the David Niblock Band of Winston-Salem, N.C.

The next outings would be the first weekend of October, when the N&W ran three of the Walton turns (shades of the J testing in 1945!) for N&W retirees and family members. Bob Claytor ran the 611 up grade to the Christiansburg summit and down into Walton, but made his way through the train meeting and chatting with the riders on each of the trips. It would be hard to imagine some CEOs doing such a thing. According to his son Preston, it might have appeared that he was play-

The morning of October 19th finds the 611 putting on a dramatic show at Christiansburg, Va., on the way to Bluefield to test the turning capability of the wye track before the public (and CEO!) were present. *Kenneth L. Miller photo*

You can just hear that exhaust barking and feel the ground trembling as the CEO backs 611 and train for a photo run at Elkhorn Tunnel near Maybeury, W. Va. *David R. Hannah photo*

ing while he was actually working. Very much a people person who loved to get out among the rank-and-file employees, he might extract a grassroots answer from someone like engineer Vance Handy as to why No. 86, with hot Ford parts, was being delayed in Roanoke Terminal. Usually, a chagrined superintendent was squirming just out of ear shot...

After the debacle in Norfolk, the operating people were taking no chances for the next big trips, two Roanoke-Bluefield & return specials on the weekend of October 23-24. The 611 ran west with tool car, caboose and some empty hoppers on Tuesday, October 19th to Bluefield, to make sure that she could turn on the wye there. They also tested further west on the mainline to Iaeger, W.Va. and turned there as well, since that trip was slated for the following weekend of operation.

Everything seemed to work fine and 611 and crew returned home on Thursday. The chilly weather on Tuesday morning provided some excellent photographic opportunities; since the move was a deadhead, not many people would witness the grand event.

The 611 performed well for the remainder of the season, not venturing off the N&W rails for this year. After a Thanksgiving weekend trip dubbed the *Oyster Bowl Special* to Norfolk and back, the 611 was stored at the Roanoke Shops. This was the only occasion the 611 would "winter" in her home town while in excursion service.

The year 1983 was 611's first full season in the NS Steam Program, beginning with a Pocahontas Coalfields Centennial trip to Norfolk in March. The rest of season, she embarked upon what would become somewhat of a standard pattern, crisscrossing the system. The longest trip scheduled was from Richmond to Chicago, after the NRHS National Convention in late July. Upon arrival in Portsmouth, Ohio, at the end of the second day, problems in the arch tubes began to show up. The fire was dropped and steam crew member Charlie Schlotthober crawled into the still hot firebox to weld up the leaks during the night. By fall, the problem was worse and 611 had to be returned to Birmingham and pulled from service until the tubes could be replaced.

Moving ahead to 1984, a full schedule again was penciled in, with a healthy locomotive. The big event for the year was her dedication as an American Society of Mechanical Engineers (ASME) National Mechanical Engineering Landmark. For the event, the 611 was under steam in Roanoke on May 19th. After a luncheon for invitees at the Hotel Roanoke, the guests walked across the street to the passenger station for several speeches and acceptance of the designation.

Come 1985, another big year for the NS Steam Program was celebrated, as another N&W locomotive was selected to join the fleet. Class A No. 1218 was pulled from the Virginia Museum of Transportation in May and shipped to Birmingham to be rebuilt to join the 611 in service. While 611 continued as the major power for the excursion trains, Southern's 2-8-2 No. 4501 occasionally filled in that year.

A highlight of 1985 happened on August 17, when 98-year-old Walter T. Carter rode the 611 from Kenova to Portsmouth on an Ohio State Fair trip to Columbus. A *Powhatan Arrow* engineer when he retired in 1957, Carter, who transferred his seniority west in 1910, had been a fellow fireman on the Radford Division with men that Road Foreman Engines-Steam Frank Collins came up under. The occasion was the presentation of a 40-year pin from the Brotherhood of Locomotive Engineers and the cab was aglow as the two talked about mutual old friends.

Ninety-eight year old Walter Carter (left) talks with Road Foreman Engines-Steam Frank Collins in the cab of 611.
Tim Hensley photo

Disaster in the Swamp

It seemed like just another excursion season until about 1:00 pm on Sunday, May 18, 1986. The 611 was running at a good (and legal) 58 mph heading westbound at Juniper, Va. about 14 miles west of Norfolk, when something went horribly wrong as the train passed over a switch. Seven cars back, things began to come undone as one car, then another jumped the track and separated from the head end. Bob Claytor, running the 611, as was his practice, saw the cloud of dust and debris as brakes went into emergency. Cars were twisted, turned and mangled. Three went onto their sides, stopping crosswise on the tracks. The 611 and first six cars were not harmed; fourteen cars had left the track. The passengers were NS employees, their families and friends, on a pleasant, simple Sunday outing to Petersburg and back.

Defective maintenance and inspection of the switch was found to be the cause. The next thing was to get the injured to care; many were air lifted aboard the "Nightingale" from Norfolk General Hospital. The remote location was a hindrance to first aid, but the company responded well and got help in the best way possible. The wreck was made worse by the fact that many older, ex-Southern cars did not have tightlock couplers, which might have prevented much of the jackknifing and overturning,

The wreck, of course, made national news. An FRA inspector was quoted in the Norfolk paper saying the injuries were more severe because of the "antique wooden cars." The wire services picked up the story and sent it out across the country. Obviously, the media did not have a clue, and the story was repeated several times over. The last wooden cars left the N&W in the early 1930s. It was up to co-author Ken Miller to call the Associated Press and explain the inaccuracy. A correction was finally sent out later in the day, but in typical media fashion, it was buried with little, if any, notice.

Emotionally distraught, Bob Claytor disgustedly sat down on a pile of crossties. Road Foreman Frank Collins assured him that it was not his fault and tried to bolster his spirits by telling him, "Now Mister Claytor, you've got to shoulder the same responsibilities as any other engineer."

An inexperienced engineer, Claytor was not. For six weeks during the long Clerk's Strike of 1978, he took Time Freight No. 86 out of Roanoke to Norfolk about six of an evening, signed off for the required eight-hours rest, then returned with a hopper train the next day. Some of his runs were made in just under six hours for the 250-mile trip.

On a Walton turn trip with the 611, co-author Hensley, himself an engineer and riding back in the baggage car, was particularly impressed with Bob Claytor's train-handling abilities. As he started east, a big horse with a female rider bolted and fell along the track ahead; the cool NS chairman brought the train to a safe, slack-stretched stop.

Before federal certification, questions arose during this critical time about Claytor's credentials. Hensley, working as associate editor and public relations representative with the Brotherhood of Locomotive Engineers in Cleveland at the time, was at the elbow of President John F. Sytsma when he made a personal call to the NS chairman and offered to vouch for his qualifications. Claytor thanked him, but respectfully declined, fearing it would hurt the labor organization leader with his membership. Simply put, it was a mutual class act between two very decent human beings.

Rightfully so, the railfan community was concerned. Would this wreck (which possibly would have occurred anyway, with the first train that passed over the spot) cause a panic and end the NS excursion program, or would Claytor hold out and continue? Fortunately, wisdom prevailed; and with only one weekend canceled to arrange for replacement equipment, the 611 continued in service. When the 611 returned to Roanoke only a day later, one couldn't have turned around at Shaffers Crossing without bumping into a white-helmeted official. At the Norfolk headquarters on the first business day after the wreck, Bob Claytor personally went around to every employee and apologized, even though he was exonerated of any blame.

Several things came out of the wreck, good or bad. First, it was the end of the Southern cars, particularly the open cars, which had served faithfully since the 1960s. Since none had been equipped with tightlocks and suffered most of the damage, they were retired and sold. A speed restriction of 40 m.p.h. was placed on all excursion trains, the 611 could no longer get up and run the way she had been designed. The corporate thinking was that slower speeds lessened the liability.

The tightlock couplers were also a factor in the 611 wreck in January 1956. Usually, head end cars were not equipped with these couplers and therefore took the brunt of the damage in the Cedar mishap. Most older cars did not come equipped with these devices and thus were vulnerable when a train derailed.

The 611 would finish the 1986 season quietly, with no further incidents. The only other real event of the year was Bob Claytor's retirement, Bob remained on the NS Board of Directors, but was no longer at the helm. Former Southern man Arnold McKinnon was now in the lead. McKinnon was not a big fan of the steam program, but did not try to kill it as long as it was not costing the company too much money.

A Reunion in Roanoke

The 1218, first anticipated in mid-1986, then by the end of the year 1986, was finally ready to come in service by April 1987. The 1218 had a lot more work to be done than originally believed and still needed some tune-up before successfully running a regular season.

The big event would be the third National Railway Historical Society Convention sponsored by Roanoke Chapter, which would be held at the end of July. It would be the first time two live N&W locomotives would be under steam together since May of 1960.

The much anticipated convention was one of the biggest in the history of the NRHS, with a large number in attendance and many more sightseers along as well. The grand finale of the event was on Saturday, August 1, when the 611 headed up the excursion train loaded with attendees, destined west from the downtown passenger station. The Roanoke Chapter had kept the event in as much secrecy as possible without letting out the pertinent details. The 1218 was waiting at Shaffers Crossing with a hopper train; the 611 headed on to Salem, where a runby location had been mowed down and a street closed off to allow the event. The fans unloaded and the 611 backed off about a mile to the east; the 1218 was heading west with her hopper train simultaneously. They paced by the runby scene, side-by-side, coordinating the move together on the double track. The 611 slowed and stopped to reload the passengers, everyone was happy. If that had been the "grand finale," everyone still would have gone home satisfied.

But that was only the beginning. The riders gave an ovation to the organizers and planners of this event when announced in the coaches what would happen next.

Continuing westward, the 611 overtook the 1218 beyond Singer – location of one of the famous company photos of the *Powhatan Arrow*. A few miles further, on the side of the mountain just west of Elliston, the passengers unloaded again; this time both locomotives were working as they climbed the grade. Reloading again, the 611 showed the great power of the Class J and what they had been designed for: Starting a heavy train on a grade, efficiently.

Catching up with the 1218 again, near Montgomery, the two trains ran up the mountain together to Christiansburg, an event never before attempted and never duplicated with two steam locomotives running side-by-side on their *original* railroad! Bob Claytor was running 1218 on the normal Westbound track; Frank Collins was running the 611 on the Eastbound.

Back and forth, the two jockeyed for position. The engine crews were enjoying the event as well as the passengers, joyfully celebrating the occasion with a water fight by punching holes in the lids of little plastic bottles squeezed like squirt guns!

As the two trains approached Walton, the 1218 slowed to a stop and allowed the 611 to use the Bristol leg of the wye and proceed on to Radford, where lunch was planned. After unloading, the 1218 came into town and dropped its empty hopper consist. Coupling to the 611, they both backed the mile to Walton and turned on the mainline wye, then backed into Radford again, loaded and returned the high-spirited train to Roanoke.

The photos truly can never do this event justice, it was something to be seen and heard. 611 and 1218 at Salem, Va. *Kenneth L. Miller photo*

Norfolk Southern has to be credited for closing down their mainline for most of the day to allow a re-creation of the past. The original idea of the side-by-side running was suggested almost in jest, knowing the unlikely possibility of the railroad's cooperation to make this event happen. However it did happen and was an event that likely will never be repeated, at least not with the accuracy and scale of this one. It was one of the most dramatic and most positive events of the 611's long career.

After the significant week in Roanoke, the 611 went one way and the 1218 the opposite; they had different schedules to make and in-

It is not always locomotives that star in the show. There are people behind them as well (see NS Steam People sidebar on pages 74-75); Jim Bistline (far left) managed the program from its formative days in the mid 1960s until his retirement at the end of the 1987 season. Carl Jensen (in a typical pose next to 1218) capably took the reigns of the program and managed it almost to the end in 1994.

In addition to the steam stars, engine and shop crews, a major factor in the success of the steam program is vested with the riders (bottom left) and the volunteers who planned, promoted and executed the trips and provided equipment to ride behind the big locomotives. Frequently, cars and volunteers were from NRHS chapters across the system. Those volunteers often provided a first, and sometimes only, train ride to folks residing in the hundreds of small towns that line the routes. Those towns may not have seen a passenger train in decades. Long time Roanoke Chapter members Elbert Miller (left) and Bill Carson both served faithfully with the steam program throughout the 1980s and 1990s and provided great PR value for the railroad. Sadly, both are no longer with us.

All: Kenneth L. Miller photos

dividual services to render. However, nothing could compare to that magical day on the Radford Division, just a tiny piece of the past was recaptured in perfection.

After the 1987 season, Jim Bistline retired and was replaced by long-time Claytor associate Carl Jensen, who lead the program through its last years. Carl had been involved in steam excursions and passenger equipment since the early 1960s and was one of the key movers and shakers to bring the NRHS Roanoke chapter back to life. Carl said about the program: "Mister Bistline was the person behind most of the actual planning of the Southern Steam Program, especially after the first couple of years when Graham delegated it to Jim to arrange and execute. He was a major factor in the Southeastern states for his railroad, railfanning and NRHS growth. The Claytor brothers were the authority persons that said do it, but the facilitator was Jim, and I inherited a pretty good, going operation. While the various steam engines became the stars of the program over the years, ending with 611 and 1218 as a duet of the best, all needed a train to pull and someplace to go. And that is what the name 'steam program' really means to me. The word 'program' defines it well, a combination of factors that are put together to deliver the product — the Southern and then Norfolk Southern Steam Program." Specifically, Carl stepped up the maintenance plan to eliminate on the road breakdowns.

611 snaking around the "S" curves climbing towards Bluefield. *Both: Tim Hensley photos*

With the famous Palisades of the New River in the background, the 611 is headed east to Roanoke on the former Virginian track at Eggleston.

The 1989 season was another one for the record books, the 611 performed admirably, just like the Js in regular service, seldom a problem arose that sent them to the shop. The biggest event for the 611 that season was the little-publicized deadhead move from Roanoke to Lynchburg. For a few days in July, there were three live steam locomotives at Shaffers Crossing for the first time since 1959. The three: 611, 1218 and Nickel Plate Road 2-8-2 587, would triplehead in the move, before splitting up to get in position for the 1989 NRHS convention in Asheville, N.C. The 611 would go in the Birmingham Steam Shop that winter for flue renewal and a general overhaul; the 1218 would handle most of the program for 1990 and part of 1991.

During the latter year, the NRHS convention was held at Huntington, W.Va. Corporately, Bob Claytor was on the BOD of Ashland Coal (one of America's largest, "magic mineral" exporters at the time), which was headquartered in the "Jewel City."

Through son Preston who was a CSX road foreman/trainmaster, co-author Hensley, by then resident vice president-WV for the competing railroad, became good friends with the father. Only wanting to attend the annual meeting and dinner when he would come to town, the two would often visit aboard the *Claytor Lake* when it was parked on the shop tracks at nearby Kenova. They would visit NS rail-to-river and other facilities in the region and even the CSX Huntington Locomotive Shops.

Upon returning to Kenova one evening, they encountered NS Car Foreman Clifford Hunt, a big, tobacco-chewing, bib overall clad, Appalachian stereotype who was originally from Williamson. Hensley electronically rolled down the passenger-seat window and introduced him to his former chairman; taken aback and visibly shaken, Hunt nearly swallowed his cud!

As part of the Huntington NRHS Convention, the 1218 powered a trip from Kenova to Iaeger and return on Wednesday, August 7. Waiting aboard his private car in the McDowell County town was Bob Claytor. A trainmaster was sent to find Hensley who, as a CSX liaison, helped coordinate the convention. With *Trains* editor Kevin Keefe (who had never met the rail industry legend) in tow, a nice visit was had aboard the *Claytor Lake*. During an impromptu interview, Claytor unhesitatingly told Keefe that his proudest accomplishments were "The NS merger and the steam program."

After the visit, the two rail buffs were invited to tag along as Claytor climbed into the cab and gingerly backed the big 2-6-6-4 and her train around the east leg of the wye. Three short toots issued forth from the melancholy whistle and rolled off the high, Tug Valley mountains. Nursing advanced prostate cancer and in obvious back pain, Claytor may have touched the throttle of one of his beloved locomotives for the last time...

Also, in 1991 Norfolk Southern marked the 25th Anniversary of the steam program and a suitable celebration was held in Chattanooga on November 2-3. Many people involved with the program over the years returned for the festivities. The 611, 1218 and 4501 were all on hand for the event: the 611 had brought a train in from Atlanta; the 1218 was running out its flue time and would be placed in the shops that winter.

Special guests at the event were Graham and Bob Claytor, as well as Jim Bistline, Paul Merriman and many others. In a rather sad statement, no one representing NS high office was in attendance. It would be the last time that Graham and Bob would publicly be together to enjoy what they had accomplished.

Bob Claytor passed away on April 9, 1993; in his honor, No. 611 flew black flags all that month. Before he ebbed into eternity, Bob Churchill presented him with one of the coveted, new federal locomotive engineer's certification cards.

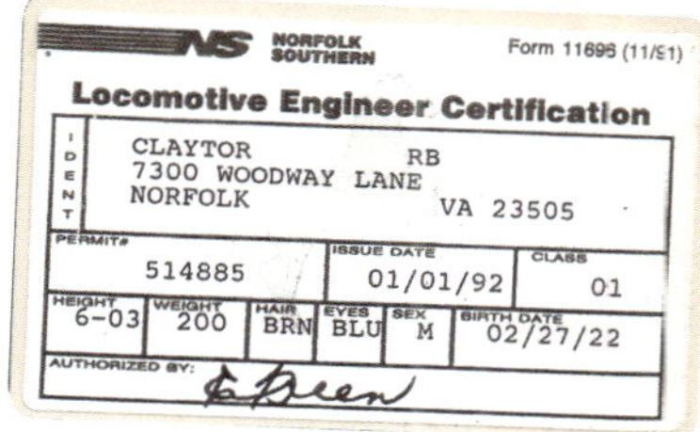
NS NORFOLK SOUTHERN Form 11696 (11/91)

Locomotive Engineer Certification

IDENT
CLAYTOR RB
7300 WOODWAY LANE
NORFOLK VA 23505

PERMIT#	ISSUE DATE	CLASS
514885	01/01/92	01

HEIGHT	WEIGHT	HAIR	EYES	SEX	BIRTH DATE
6-03	200	BRN	BLU	M	02/27/22

AUTHORIZED BY:

Preston Claytor Collection

Frank Collins passed away in 1990, Jim Bistline had retired in 1987 (He succumbed at age 90 in 2005.) — the old guard was disappearing fast. Regretfully, there was no one in upper NS management that would be considered a friend of the program.

In 1993, the 611 handled the program with no difficulty, while the 1218 was still undergoing work. It was a season with a variety of trips and 611 running lots of miles. The program's trains were now more reliable and, with modernized mechanical systems and a completely matching train set, had received lots of good comments. Looking ahead, the train should have been be good for five years of service without major work.

Bob Claytor, hand on the throttle of the 1218.
Kenneth L. Miller photo

Things are not well...

The 1994 season was going fine, but things behind the scenes were not good. Graham Claytor, original patron of the steam program passed away on May 14. About a week later, upper management ordered the Birmingham shop crew to stop work on the 1218. The 611 was still out on the road and would cross the system again that season. The J was due for shopping after the end of the year (The flue time would expire in January of 1995.), and she would require either an extension or replacement of the tubes. New flues were already on hand, but with work stopped on the 1218, it would mean a big push on the 611 to have it ready by the end of March.

In September, one of the long deadhead moves was underway. With no available crane or loader in Roanoke, the 611 and train would have to go all the way to Lynchburg to get coal.

Late that night, a piggyback train was backing a set-off into Kinney Yard, and was misdirected into the wrong track and slammed into the rear of the parked-for-the-night excursion train. It was not like a derailment at 40 mph, but the damage was severe. The 611 received no damage, but several cars were damaged so badly they were deemed not repairable. After the cars were rerailed, the 611 moved the hospital train back to Roanoke the next day. A number of cars took damage, but the rest still had to be inspected carefully. That weekend's trips were cancelled, but replacements were quickly arranged and brought in to finish out the season.

On Saturday, October 30, 1994, the 611 was bound for Roanoke just east of Vicker, Va. This would be the last excursion into or out of Roanoke prior to the end of the program. The news had been released the previous night just before 5:00 p.m. and many fans along the way had not yet heard. *Kenneth L. Miller photo*

The 611 ran a Roanoke-to-Bluefield trip in early October. Then running on previously untouched territory, handling the first leg of a Roanoke – Glasgow – Lynchburg – Roanoke trip, the 611 would turn on the wye track with the C&O, then run back home "lite."

Among some observers, there was a feeling that this might be the last of 611's under steam trips from Roanoke. In fact, several years before, the possibility of the final trips had been discussed and how to tastefully end the program. It was believed that a Roanoke-Bluefield trip would be as close to the same ending as in 1959 that could be done. It was never imagined that the fans would not get the chance to close it out with a proper send off.

One last turn and service at Shaffers Crossing, October 30, 1994. *Kenneth L. Miller photo*

Apparently, there was some NS discussion that the steam program would be cut back for use on limited special occasions and that 611 would be kept in the surviving "long house" at Shaffers Crossing. This was to have been accomplished under the responsibility of Assistant Manager-Mechanical Engineering Tommy A. Aker, but never came to fruition.

The bomb dropped just before 5 p.m. on Friday, October 29. The program would end after the last trip scheduled out of Birmingham in early December. No ceremony, no event, no grand homecoming for one last trip — just quit and hope it goes away without notice.

Waves of shock rolled through the railfan community; word had been out for several days before among the inner circle. Norfolk Southern management even hoped to get by with no news release. However, they did the next best thing they could, by sending out the release just before 5 p.m. on a Friday, then disappearing for the weekend. The 611 was running from Bristol to Roanoke and return the next day and almost as many people came out to see her come in as did on that warm August day in 1982. But, this time there would be no speeches, no ceremony, simply many folks who could not believe the news that was in their newspaper that morning. A number were upset and angry; the out-of-town fans chasing had not heard the news. Even though, the roundhouse was now gone, the 611 came to Shaffers Crossing one last time and then was sent back out, servicing and turning in the old style again, proving the high availability and reliability of the Class J.

Efforts, including a Save Our Steamtrain ("SOS") grass-roots campaign, were made to try and convince NS that this could be worked out to continue the 611 excursions without the company's investment. Management was not at all interested. They politely listened, but publicly compared the 611 and her vintage cars to a wagon train on the present interstate highway system; it simply did not fit on a modern railroad.

In fairness to NS, traffic levels were increasing since the 1980 passage of the Staggers Act, which deregulated the railroads. Because of a new-found rate freedom, coal, merchandise and especially intermodal movements would grow to record levels and an everyday steam train would just get in the way.

How they handled it (ultimately, the decision of Chairman David R. Goode – who was likely influenced by Vice President Thomas J. Golian, Director-Staff Services Robert A. Lovelace and Executive Vice President-Operations Paul Rudder and his successor Stephen C. Tobias) will not go down in history as a case-study, textbook example of how to promote positive public relations.

The final steam trip was running on December 3, 1994, from Birmingham to Chattanooga and return. Chattanooga was where the Southern steam program had all started some 28 years before, now it was getting a sad honor of being the last public destination.

Special guests from all over were there for the train, including a number of representatives of NRHS chapters and other trip sponsors. It was an emotional trip, riding Tool Car, No. 1407 for the last miles into town. Many of the faithful were on hand to chase and see the last trip, including Jim Bistline, former manager of steam operations. The Tennessee Valley Railroad Museum had 2-8-0 No. 610 out to pose with the 611, but it was hardly a celebration. Special commemorative white flags had been made to place on 611 for the trip. The weather was, fittingly, gloomy and cloudy for the day. On return to Birmingham, the skies welled up and cried, as a heavy rainstorm

It is another dark day in history: December 7, 1994. Co-author Tim Hensley (white hat) speaks with Bob Saxtan at Lynchburg. The lady in red wiping her eyes is Jane Hicks, who was secretary to the steam program. *J.L. Sanders photo*

pounded down as the passengers were unloading for a final time.

The 1218 was sitting in the shop at Birmingham, cold, mostly disassembled and looking forlorn. The 611 would return that night and join her, but not for long.

The 611 and train would leave Birmingham very early on Monday morning, the 5th, headed for Atlanta and then on to Salisbury and finally into Roanoke. It almost seemed that there was a feeling among the company that this was something to be ashamed of and everything was done to keep the move quiet. On December 7 – Pearl Harbor Day again, 1994, N&W 611 returned, under her own power, to Roanoke for her second –and what appeared to be final – retirement. Fittingly, the day was overcast gray, not cold, much like that storied Sunday fifty-three years before. The mood and lack of understanding seemed to match that "Day of Infamy."

It is a gloomy day at Chattanooga, as TVRM 2-8-0 610 is out to greet the soon to be retired 611, just in from her final public excursion. Note the commemorative white flags flying on the 611. *Kenneth L. Miller photo*

Three faithful friends – Ken Miller, Jeff Sanders and Tim Hensley – drove over to Lynchburg to escort her home. This time, there was no eager group to wash the locomotive so that she would look good upon her last arrival, no dignitaries, just a sad steam crew rolling away the final 52 miles.

Before departing Lynchburg, the steam crew had placed a Biblical sign on side of 611's cab that read: "Forgive them for they know not what they do. The Thoroughbred of Steam, meant to run free." An NS official had it removed before they left town.

Along the way, many others joined the Route 460 funeral procession. They stopped in Bedford to watch her by. While Miller and Sanders solemnly snapped pictures, Hensley just wanted to watch and indelibly etch the images to eternal memory.

Once she arrived in Roanoke, the three amigos were waiting on the Firth Street Bridge. Looking down, a Norfolk Southern officer guarded the scene, as switches were lined into the terminal. A former company official, now deceased, had made an ill-advised warning to the railroad's management that there would be trouble, speculating that there would be public protests or even trespassers emotionally throwing themselves across the tracks! *Now why in the world would anyone want to mar the final tribute to something they loved so much?*

When the locomotive moved out, it made a wincing sound as it ambled west. It was a cry that no one ever heard her make before, perhaps somewhat akin to a death rattle. As she passed underneath, by then Road Foreman of Engines – Steam Bobby Saxtan – dutifully sitting at her throttle for one last time – removed his glasses and wiped his eyes. He wasn't the only grown man in Roanoke to shed tears that day...

She moved almost silently, no grand salute like when she arrived in 1982, no flags, no bunting. She looked a bit tired and ready for the drama to be over, not unlike 1959. Unnecessary whistling had been banned; officials did not want to draw attention to her plight. But, one long blast issued forth as a last breath to immortalize her forever.

That evening, after supper with his longtime friends, Hensley stole into the Shaffers Crossing Engine Terminal, intent on having a special moment with the 611, much like an immediate family member would want before closing the lid on the casket of a loved one. In the early winter darkness, he tenderly kissed the locomotive on her right, engineer's-side cylinder while she was still coursing with steam.

It was a long, melancholy drive home to Huntington. Memories of what the "smoke engine" (an affectionate boyhood term) had meant to his life, his friendship with and respect for the Claytors and others on the steam crew swirled through his mind as he negotiated the curvy West Virginia Turnpike.

The next morning, he noticed a smudge of soot on the bill of a white, Operation Lifesaver cap he was wearing (at the time he was also West Virginia state coordinator for the rail-highway safety program). As expected, Hensley has kept that hat: a symbol of the final, physical contact he had with the beloved machine. He hasn't worn it since and never will.

Speaking of hats, co-author Miller was wearing one he wore on the trip from Birmingham to Roanoke in 1982 and would wear once again in 2015 on return to Roanoke.

Now, the steam crew would prepare 611 for long-term storage. A few days later, the locomotive, now cold and dead, was hauled to the Roanoke Shops for storage in the old Paint Shop until a new berth could be made at the Virginia Museum of Transportation, located now at the rambling, former N&W Freight Station, since the 1986 move from flood-prone Wasena Park. It would be almost 10 months before the spot was ready.

Dismantling the steam program progressed rapidly. At the end of February, a large auction was held at Birmingham and many spare or new parts for the two locomotives were sold off. Tool Car No. 1407 was even auctioned off, but in a far-sighted move, the Roanoke chapter purchased the car with a feeling it may again run behind steam. A number of the excursion train cars were also sold off at the same time. It was a depressing day; the dream was over.

In September 2007, 611 was allowed to venture out of VMT for the 100th Anniversary of Roanoke Shops and was posed on the turntable for a night photo session. Dominion Bank towers in the background. *Richard D. Shell photo*

The 611 was returned to the Virginia Museum of Transportation on October 7, 1995, to be placed under a new shed for display with one of the tool cars that sometimes followed along behind her. The former Grand Trunk lightweight car was used for a time, but was not the one that faithfully rolled up the most miles behind the locomotive.

Over the next half-dozen years, Roanoke chapter member John Garrett (like Jack White long before him), arranged for movement back and forth every few months to help keep the bearings loose and rolling – he hoped that someday soon the lady could steam again. She would make occasional "dead" outings from her comfortable incarceration at VMT, visiting the nearby Link Museum and made a return to Roanoke Shops for an anniversary celebration.

Though 611 was fittingly preserved – exactly where she should be – and positioned to be admired in awe by future generations, her plight could somewhat be likened to life without parole. Or so it seemed...

••••

Norfolk Southern Steam People

Many were associated with the Norfolk Southern Steam Program during its 12 years of existence.

Of course, much was inherited through what Bill Purdie and his Southern Ry. team had already established. Initially, the venerable SR master mechanic-steam and his son, Donald, were instrumental in overhauling the 611 and overseeing operations on some of her first trips.

Over time, the Birmingham-based mechanical program took on more of a Norfolk and Western flavor. Two people that shaped that school of thought were Frank Collins and Paul C. "Pappy" Housman.

Collins became a trusted colleague of Bob Claytor's. When the two first met, the young legal department officer would ride the head-end of through SR passenger trains from Roanoke to Pulaski, Va. During pleasant weather, Claytor would often join wife Francis and their three children (who were already there and waiting to pick him up) for getaway weekends at Claytor Lake. Graham Jr. also had a home at the modest family compound, where they kept a working steamboat docked up a cove.

A steam-promoted engineman, Collins, who was stern and serious about his work, but possessed of a wry sense of humor, often caught Y6 runs off the extra list during the last days of the big locomotives. With the Bristol Line having been officially dieselized on December 31, 1957, the coaler at Vicker, Va. became somewhat lazy, what with only an occasional, steam-drawn Radford Division "River Run" for Bluefield to service. One time, Collins got tired of being delayed, dismounted from the cab and drew a shiny, cap-buster, toy pistol on the slacker. "When those blanks went off, it liked to have scared that old man to death!," Frank later laughed.

Housman who served as Assistant Erecting Foreman at Roanoke Shops, retired from N&W in late 1974 and was brought on as a consultant. He was a walking book of knowledge on steam locomotives. Having hired on in 1925 before he was even 15, the soft-spoken Pappy worked his way up from laborer to machinist, before taking a supervisor's position in 1943. "I had total charge of the second shift," he told co-author Hensley. "I've seen as many as 75 men working on an engine under construction." Having never married, Houseman mentored the Birmingham shop folks and steam road crews like sons.

Bob Claytor and Pappy Houseman supervise liberation of 1218 from VMT in May 1985. *Kenneth L. Miller photo*

Gradually, the philosophy changed from one of "just getting the locomotives out the door" to a long-term plan of superbly maintaining the big N&W engines – similar to the way they actually were when in revenue service. "It took four or five seasons to get everything tuned up," remembered Preston Claytor, who went to Alabama fresh out of college and served as a fireman during 611's first excursion season.

"We would have two air pumps on the engine, two completed and ready and two being worked on the floor with component parts," Claytor recalled. The 1218 was undergoing the same process when the program ended.

Two that subscribed to the preventative maintenance plan were Shop Foreman Doug Karhan, and his sidekick, Bob Yuill. Using original drawings saved by former N&W Project Engineer Mark Faville, they oversaw two, overlapping 12-hour shifts, to get the 611 back in working shape. This was before she was first outshopped in 1982.

Among those who worked under their supervision were Scott Lindsay, Tim Sposoto, Greg Todd, George Walker, Boilermaker Don Pate, Al Phillips, Bill Magee and Dwight Browning. Years later, even Secretary Jane Hicks came out of the office to file burrs off rod brasses!

Once on the road, many like Browning, Walker, Webb, Chuck Jensen (son of Carl) and Lindsay worked as firemen. Others who took their turn at the stoker controls or as volunteer night watchmen (A modern take off on the old "fire & water tender" title.) included John Grubb, Sandy Alexander, Kelly Barnett, Percy

Wilkins, Jeff Sanders and John Barnett.

When working out of a yard or terminal, the called engineer on a crew was usually – at least at some point during the trip – afforded an opportunity to take the throttle. Bob Claytor always saw to it that engineer Tom Dressler, an N&W expert and one of the best modelers that ever lived (He died in 2007.), was in position to work the job up the Scioto Valley out of Portsmouth.

One who wasn't invited to take a turn was a braggart who climbed into the cab on one of the Midwestern tours. "I can't wait to beat this ole girl to death!," he authoritatively announced. Ever protective of his baby, Frank Collins turned, cocked his head, raised his eyebrow, looked up over his glasses and calmly countered, "Not this old girl." He never gave up the right-hand seat that trip...

Some of the road foremen who took an exceptional interest in taking care of the steam crew included Doug Anderson of Portsmouth, J.D. Hinkle and Guy Kelly (both old steam men) of Bluefield, John Whittaker, Freddie Kite and Gary Myers. Trainmasters Jerry Nassar, John Irwin (then at Bluefield and now NS AVP-Transportation Network) and Rick Keene at Kenova were always helpful. Air Brake Car Instructor S.R.Wineguard and Dispatcher Tommy Duncan also assisted.

Eventually, Collins trained affable, Bluefield-born, Norfolk yard conductor Jack C. Taylor, Jr. and the somewhat shy, sometimes sovereign Robert Saxtan as his replacements. Both Saxtan, from Reading, Pa. and Scott Lindsay, a New Jersey native, are "Yankee" holdovers from the early Ross Rowland days.

According to Taylor, Frank liked to let his students get themselves into a problem and then see if they could work their way out. Once, while letting one of his old Radford Division cronies fire, Collins was asked at the end of the trip, "How'd I do?"

"You couldn't fire then and you couldn't fire now!," the steam road foreman teased.

Although the envy of many, being a member of the NS steam team meant long hours and hard work. Still, some couldn't always control their excitement for such a position.

On a Williamson-Lynchburg deadhead move, Taylor remembers running east on the 611 with about 25 cars, including two full canteens (one more than normal) of water and three gondolas of coal. Engineer Clyde Barr advised that the speed at which they went over the last road crossing at North Fork, W. Va. would be about the same as what they would make up Elkhorn Mountain. Marching right along at 15 m.p.h., each slowly-measured exhaust from the stack was reporting like a shotgun blast.

With his usual cigarette dangling from his lips, fireman Steve Webb smiled, looked over at Taylor and shouted: "Awesome, Mister Jack! Just Awesome!"

••••

Clockwise, from top right: Jack Taylor (far left, red cap) and Dwight Browning (on ladder, blue cap) do an on-the-road repair to 1218. Below, Jack Taylor cleans the fire. Left bottom, Bob Saxtan is cleaning 1218 for the next run. *All: Kenneth L. Miller photos*

Part 3 : FireUp611!

Top right: The big day at last, 611 is slowly pulled out of the Virginia Museum of Transportation, May 24, 2014. Destination, Spencer, N.C. Bottom left: "Wick" Moorman removes the ceremonial first bolt to start the restoration. Bottom right: 611 is the centerpiece among a gathering of cab unit diesels called *"Streamliners at Spencer"* at North Carolina Transportation Museum, May 31, 2014. One can see exactly how carefully the 611 must be spotted on the 100-foot turntable.
All: Casey Thomason, Norfolk Southern

It is said, so goes an old Blue Ridge Mountain euphemism, that "The apple never falls far from the tree." "History repeats itself" is another adage that has long been heard throughout America.

It could be argued that those two sayings had a hand in awakening a much-admired steam locomotive from a 21-year slumber. And maybe it is fitting that a Claytor had a resuscitation team role in bringing No. 611 back to life.

Setting the stage was the June 30, 2010 announcement that "Mainline steam excursions could return to Norfolk Southern rails as early as 'this fall' after NS said that it is negotiating with Tennessee Valley Railroad Museum (TVRM) to operate a limited schedule of steam locomotive appearances and public passenger excursions beginning later this year." Naturally, a change in attitude had to take place to ignite such an ember and that spark burst into flames on February 1, 2006, the day Charles Wilson "Wick" Moorman, a self-avowed rail historian, succeeded David Goode as Chairman and CEO of the rail industry giant.

"This is the right time for steam to ride the Norfolk Southern rails," said Moorman. "We have a fascinating history and we have a compelling message about how today's railroads support jobs, competition and the economy."

Calling the rekindled program *"21st Century Steam,"* it was not until 2012 that the program got up and running, what with TVRM's ex-SR 2-8-0 No. 630 operating on a series of public excursions and NS 30th Anniversary employee specials in Virginia, North Carolina, Tennessee, Georgia and Alabama. That same year, Fort Wayne Railroad Historical Society's splendid former Nickel Plate Road 2-8-4 No. 765 joined the stable, scheduling trips in Ohio, West Virginia, Pennsylvania, Indiana, Missouri and Illinois; in addition to its own, a heritage whistle was utilized to honor portions of the NYC, PRR, N&W and Wabash predecessors over which it would pass.

Taking a little bit longer than what was originally planned, the TVRM's real focus in the participation was the restoration of fabled "poster child" Southern Mikado No. 4501. This time in black freight livery, instead of the storied Harriman green that the USRA 2-8-2 sported for her first excursion career, she made some break-in TVRM excursions in 2014 before making her first *"21st Century Steam"* appearance from Bristol to Radford and back on June 28, 2015.

With so much going on around them, the N&W die-hards soon took notice. During the 2010 TVRM 50th Anniversary banquet in December, Moorman spoke about the limited steam train that NS was putting together. Taking questions from the floor, the CEO conceded that the Virginia Museum of Transportation had two of "the crown jewels of steam." Citing that the various repositories of railroad interest in the Star City – VMT, the Roanoke Chapter NRHS, Link Museum and N&W Historical Society – were all fractionalized and competing against one another, he challenged them to consolidate their efforts. Bottom line: This was an open invitation to take a look at the 611.

For Preston Claytor, "It was like waving a red flag in front of a bull!" Perhaps inspired by staying on the *Delta Queen*, he approached Moorman after his Chattanooga speech and vowed, "I'm going to hold you to that!"

Through his friend Eugene M. "Gene" Elliott, Jr., Esq., a Roanoke attorney who is on the VMT Board and lives in the former Claytor homestead, he commenced to fan the flames. "Let's get this thing out!" he told the lawyer who is also the museum's treasurer.

Together, they convinced Beverly T. Fitzpatrick, Jr., VMT's executive director, that it was the right thing to do. Earlier, the city council had passed a resolution designating the locomotive "The Spirit of Roanoke." In 2012, the City of Roanoke had formally given title of the 611 to VMT, so there was no excuse of questionable ownership. A former vice-mayor and very civic minded, Fitzpatrick envisioned the locomotive as a rolling billboard for his hometown.

PART 3

Over the next two years cost estimates were solicited and a business plan was formulated. In the meantime, a lease proposal from Cheri George, a former volunteer fireperson on the 611 during its 2nd life and now a successful Atlanta software consultant, was considered but not executed.

In December 2012, the informal committee met with Wick Moorman to get his final blessing. "You won't lose money, but you won't make a fortune," he told them.

On February 22, 2013, VMT announced a feasibility study of returning the "iconic" Norfolk & Western steam locomotive to operating condition. Out of these initial efforts, FireUp611! was born.

Ever the politician, the astute Fitzpatrick

appointed the committee: Ron Davis, president of N&WHS; Jeff Sanders, president of the NRHS Roanoke Chapter; Ken Lanford, president of the VMT Board; Jeanne Bollendorf, executive director of the O. Winston Link Museum; William L. Harris, president of North Fork Lumber Co., Goshen, Va.; consultant Jim Stump; Jim Wrinn, editor of *Trains* magazine; Trey Davis, of the Urban Land Institute in Washington, D.C. as a marketing consultant; and William Withuhn, curator emeritus of the Smithsonian Institution. Bringing some real railroad expertise to the table were Chairman and Project Coordinator Preston Claytor of Boynton Beach, Fla. who served many years as VP-Safety & Operating Practices for the RailAmerica consortium and Scott Lindsay, resident steam expert and president of Steam Operations Corp. of Birmingham, Ala. Recently, Col. Lewis Ingles Jeffries, U.S. Army retired and author of *N&W: Giant Of Steam*, was appointed historian, replacing the much-respected Louis M. Newton who fell ill.

Scott Lindsey and Preston Claytor, pictured with a Southern unit and 611 representing the two combined railroads. *Noah Bimrose photo*

Last but not least, the aforementioned Cheri George would become the administrative glue that held the diverse group together.

A fundraising campaign was launched and nearly 2,000 donations poured in from across the nation and 15 foreign countries. Norfolk Southern stepped up to the offering plate with a $1.5M endowment. By October 25, 2013, it was announced that the Capital campaign had raised enough to restore the locomotive. "It's no longer a matter of 'if' Class J 611 will run again, but 'when' she'll return to excursion service," Fitzpatrick said.

All the while, part-time VMT employee Charles Hardy, who coordinates black history exhibits and interpretations, did his part. "If you want to get to Heaven, better give to Six-Eleven!" he told practically everyone he came in contact with...

"Once we had the numbers, we turned to railfans around the world and asked if they would help," Executive Director Fitzpatrick said. "They supported the campaign in a way that we simply could not have imagined."

With $3.5M in the bag, the museum (which won top honors with the FireUp611! campaign in the American Advertising Federation and Blue Ridge Public Relations Society's 2013 competition) turned its attention to raising enough money to build an on-site maintenance facility. The new building will house equipment to maintain the locomotive in serviceable condition, an exhibit gallery and an educational center, so students of all ages can learn about steam technology, science, engineering, math, social sciences and history. The facility will also keep the 611 accessible to the public during scheduled periods of inspections, boiler washes and general repairs.

"We believe that the facility will protect our donors' investment in the Class J 611," Fitzpatrick mused. "We want future generations to enjoy her power and elegance." In 2012, visitors from every corner of the United States and 49 foreign countries visited VMT to see the 611, 1218 and other exhibits reflecting all forms of motion.

"Roanokers should feel proud that their city built such a beautiful locomotive," the museum director touted. "She was designed and built by the craftsmen of the Roanoke Valley. She's a symbol of our past and a signal of what we, as a community, can do in the future."

Initially, a cart before the iron horse, "build it and they will come" mentality prevailed in regards to the construction of the building. But after the generous NS contribution was received, it was decided to proceed first with the restoration of the locomotive.

Returning to the immediate task at hand, arrangements were made with the 57-acre North Carolina Transportation Museum to restore the locomotive at its 37-stall Bob Julian Roundhouse in Spencer, N.C. With a 100-ft. turntable and shop, the facility boasts one of the last venues equipped with the necessary tools to restore massive locomotives.

Accordingly, VMT employed Scott Lindsay of Steam Operations Corp. to restore the 611. Up-to-date expertise was needed to meet current safety (particularly, boiler) standards and Federal Railroad Administration guidelines.

Restoration Photo Gallery

There are thousands of details in any steam restoration. Here are just some of those as depicted by Noah A. Bimrose. Clockwise from left: A newly rebuilt Westinghouse Cross-Compound air pump has arrived. Top right, when you have no suitable overhead crane, you improvise. Bottom right: the backhead receives attention. Bottom left: Fitting up the boiler jacket.

During the last year of the NS Steam Program, Lindsay's company (a "traveling steam road show," as he calls it) had a contract to maintain the two former N&W locomotives. Among the engines he has worked on are: Nickel Plate No. 765; Spokane, Portland & Seattle 4-8-4 No. 700; Frisco 4-8-2 No. 1522; and some smaller Nevada Northern engines. He has also done consultant work in Canada and Ecuador. More recently, he mastered a towed move of C&O No. 614 from New Jersey to Clifton Forge, Va. (via Roanoke!) for Jim Justice's stillborn *"Greenbrier Express."*

Joining the primary VMT Mechanical Team were: Doug Karhan, chief mechanical officer, emeritus (former NS Master Mechanic-Steam in Birmingham); Robert Yuill, boiler specialist (another Birmingham vet); and Robert Saxtan, former NS steam engineer and now senior general foreman who would address safety appliance, airbrake and support system issues, aside from serving as a liaison to operators of *21st Century Steam* trips. Lindsay himself would serve as project leader and running gear specialist, assisted by General Foreman Tom Mayer.

Prior to her leaving for Spencer, a gala, Memorial Day weekend send-off was held at VMT on May 24, 2014. Five thousand attended the event, including seven descendents of Engineer N.B. "Nebbie" Gordon of Bluefield; in his memory, they proudly wore matching, blue denim shirts proclaiming their heritage on the back. Another giddy celebrant was Lois Ponton who sported a hand-stitched, customized purse, created in the likeness of 611's nose.

Promptly at 2 p.m., NS diesels coupled up to her and pulled her backward on the first leg of a long journey to a third career. Like a giant Pied Piper, she was faithfully followed by hundreds as chaperones to her current rechristening. Cheers and applause could be heard for blocks away!

After servicing and mechanical preparations at her old N&W Shaffers Crossing haunts, the 611 was towed east over the former Virginian via Altavista, Va., thence down the old SR main to NCMT. She arrived on May 26.

More good news came on May 28, when it was announced that a five-year lease had been worked out with The Museum of Transportation, St. Louis to bring Y6a 2-8-8-2 No. 2156 back to VMT, where she would finally be reunited with her "A" and "J" sisters.

The big Class J came into town just in time to participate in the *"Streamliners at Spencer"* event that was held May 29 - June 1, 2014. Centered on the turntable, she took part in a special photo opportunity, surrounded by 25 diesel cab units painted in the original color schemes of predecessor railroads. More than 9,250 tickets were sold to fans who attended the four-day event.

Taking part in the festivities was NS Chairman Moorman who turned a ceremonial first wrench on the project. "From the smallest acorn, the tallest oaks sometimes grow," so it was that an insignificant nut was officially removed from the sleek Class J's skirting on Friday, May 30. A genuine Stillson pipe wrench that belonged to James S. Stump, Sr. who worked for N&W at the Roanoke Coach Yard from 1934-1975 and was the father of FireUp611! Committee Member Jim Stump, was used for the task.

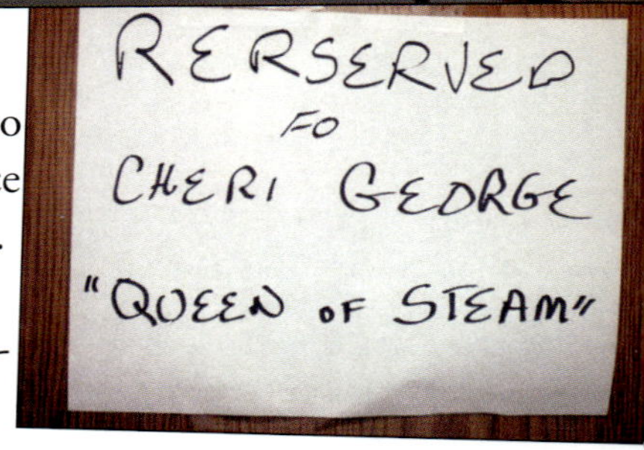

Top: Noah Bimrose photo, Bottom: Tim Hensley

A VMT trailer was moved on site to serve as an office and crew room. From her seat in the rudimentary structure, Cheri George would coordinate more than 10,000 hours of volunteer and other labor (see sidebar on pages 87-88).

By July 1, most of the disassembly was complete. In short order, the front end and superheaters were removed, and the old flues cut out. The streamlined jacket and insulating lagging were removed, the boiler shell needle-scaled and 20 years of accumulated rust removed from the interior and tender. Both Westinghouse air pumps were removed and sent off for evaluation and grates extracted for repair. Work also commenced on the lapping of steam delivery pipes, a true – almost lost – restoration art accomplished with a special compound.

Next, the engine and tender were separated (July 12, 2014), the boiler interior sandblasted and the locomotive was moved inside the roundhouse for a boiler survey. In accordance with federal regulations that call for the process every five years (technically, 1,472-days), the long, arduous task of removing, inspecting and testing all flexible staybolts was started. Also, the stoker screw and feedwater pumps were addressed. And in keeping with a Class J's correct appearance, the cab valve handles, throttle and injector (plus their quadrants), hand rail stanchions, builders plates and class plates were sent out to be re-chromed.

While everything was ripped out of the boiler, an ultrasound was performed in early August and it took several weeks for the test results to come back. Literally, this was necessary to see whether the boiler was thick enough to be safely steamed again. If the metal had been

New flues are laid out on the bench while 611 waits in the background. *Tim Hensley photo*

found to been too thin, a plan of strengthening repair work would have been developed.

At that point, repairs had the cushion of being a couple of weeks ahead of schedule.

Bob Yuill at work rebuilding the safety valves. *Tim Hensley photo*

Meanwhile, cylinder heads and piston valves were removed and their openings covered with cardboard to keep dirt out. Quietly carrying on work in the expansive roundhouse background was the bearded, bespectacled Bob Yuill who repaired and neutralized (to arrest rust) Elesco Type E superheaters. On October 24, a flatcar of flues arrived on the premises.

Over the late fall and impending winter, the heaviest of 611's second restoration took place. Most difficult of the work – and what would prove to be for the entire project – was the replacement of the rear flue sheet. "The drawings did not match what had been on the locomotive," Chief Mechanical Officer Scott Lindsay

Installing flues. Note cardboard covers over the cylinders. *Steve Morgan photo*

said. "Gary Bensmen [of boiler contractor Diversified Rail Services] caught the mistake just before he was about to cut a new one in Chattanooga," Lindsay disclosed.

Another minor problem was when the exhaust stand (upon which the nozzle sits) cracked and had to be specialty welded. In an effort to expedite and remedy that problem, "Preston Claytor had to make a hurried automobile trip to Buffalo, N.Y.," Lindsay revealed.

Though she was far from her birthplace, the 611 never got homesick while in Spencer.

Stoker motor with the star hanging from under the cab. *Tim Hensley photo*

That's because someone thoughtfully hung a red miniature of Roanoke's Mill Mountain Star under her cab!

Once Bensman and the congenial Steve Morgan of AES Manufacture (a recommended

TVRM vendor from Chattanooga) had expertly welded the new flue sheet in, it was just a matter of putting the 611 back together. That significant action was a halfway benchmark to the project.

The first new flue was threaded through the front sheet at 2:40 p.m. on November 20, 2014. The last was rolled at 2:20 p.m. on Sunday, December 14.

Above: The new rear flue sheet is being prepped for installation. Below right: Rebuilt or new superheater units are now installed. *Steve Morgan photo*

Over the holidays, a two-week break was taken by FireUp611! Many of those involved hadn't been home for months.

Upon their return, the boiler was sealed up and made ready for a 375-lb. hydrostatic test. Most of January and February of the new year was spent "chasing leaks." If one was found, the faulty staybolt had to be reexamined and, if necessary, the cap removed, bead blasted and

New staybolt caps. Bottom three: *Steve Morgan photos*

its pressure-bearing seat ground. Sometimes, the problem was as simple as installing a new copper washer. The 611 passed this important hurdle with flying colors on February 23, 2015.

While the final touches of putting the 611 back together in Spencer were taking place, the VMT support staff busied themselves back in Roanoke. Director of Development Fran

An assortment of the vintage tools. *Steve Morgan photo*

Ferguson continued fundraising and long-term planning efforts, while Peg McGuire, director of Marketing & Communications, focused on promoting the locomotive, answering inquiries about its return and organizing bus trip visitations and shop tours. Gary Gray was hired as rail transport coordinator to work as a liaison and to assist with ticketing efforts. In addition to carrying on their normal, daily VMT functions, all pitched in to plan an appropriate event worthy of the big Class J's return.

VMT's 611 gift shop efforts, lead by Susan Loveman, included the production of *Dash Dupree and the Queen of Steam,* a children's book written by Alice Standish and illustrated by Ann Glover. It is available for $16.99 in the museum's gift shop or on-line by visting *www.vmt.org.*

Bob Yuill and Scott Lindsay discuss an issue. *Noah A. Bimrose photo*

Work in Spencer reached a frenetic pace as the big day approached; shifts were increased to two a day. Safety valves (Bob Yuill), the throttle linkage and cams were serviced. New washout plugs and some sleeves were applied. A second dynamo (steam-driven electric generator) was purchased to support appliances not required when 611 last ran in 1994. Gauges were serviced, tested and installed.

Finally, the moment of truth arrived. For the first time in more than 20 years, No. 611 felt fire on her grates on Tuesday, March 31, 2015. The Internet was humming with on-scene reports as the hand on her steam gauge slowly inched toward 300!

While a welcome column of coal smoke brushed the Carolina blue sky, necessary support work had been going on elsewhere at NCMT. The week before, the tender was painted by Chris McBride of Freight Car America's Roanoke facility (His employer donated 40 hours — He actually worked 84 and then took a week's leave of absence to finish the job!). Co-author Ken Miller designed the replacement stenciling for the famous "Norfolk And Western" lettering on the side and spearheaded a team of Roanoke volunteers with the application.

Axalta Coating Systems LLC generously donated Imron paint for the project. It took 30 gallons of black, eight of Tuscan red and four of gold, aside from product required for mixing. Also, 24 gallons of clear were applied as a topcoat, giving the 611 her desired glossy appearance. Willetts Rail Car Services prepared the tender surfaces and constructed a paint booth in a confined roundhouse stall.

New transverse arch tubes. The 611 was the only Class J so equipped. *Steve Morgan photo*

The FireUp611! project attracted significant national media attention. CNN *Money* profiled Scott Lindsay and the mechanical team on February 2. In the lead comment of the April 2015 issue of *Railway Age*, Editor William C. Vantuona devoted a whole page to the effort. When was the last time *that* trade publication focused on a steam locomotive?

Greasing the crank pin. *Noah A. Bimrose photo*

The real moment of truth, Scott Lindsey pulls the throttle out for the first time. *Noah A. Bimrose photo*

Vantuono's slant was primarily on the rail industry suppliers and their donations, particularly Graham-White, a Roanoke Valley corporate citizen that has been in business for 116 years. The sanders long-marketed by the company were developed by M. L. Fowler, a Radford Division engineer and two former employees who withdrew from N&W service to establish the firm, wreckmaster John F. Graham and Roanoke Shops pipefitter W.H. White.

"Our locomotive is a rolling showcase of Graham-White parts," said Preston Claytor in the article. Included on the 611 are air brake valves, air driers ("spitter" valves, to remove moisture formed by heat during compression), "Salem" brand sanders and even the bell valve. "We did decline their offer of one of the new electronic bells," Claytor laughed.

An unplanned, historic first move under steam was made on Saturday, May 9, when a ton of sand was added to the dome for the testing of appliances. Like the dents on the tender, the last vestiges of the Cedar wreck were smoothed from that receptacle during the Spencer repairs. To the delight of those attending National Train Day at NCMT, roaring, sauntering, steam from cylinder cocks again pierced the air, while three, short "walking speed" moves were made.

As a finishing touch, Scott Lindsay set the valves on the Baker mechanism (his specialty) and painted other parts of the running gear. Looking back over the year-long project (which by normal circumstances should have taken two), Lindsay said the most pleasant surprise was "how well all the calculations worked out." This, he attributes to the expertise of Mark Ray, engineering manager of his Steam Operations Corp. In summation, Lindsay said, "Everyone worked well together."

The VMT Board had voted to officially name the 611 "The Spirit of Roanoke." On Friday, May 15, Executive Director Bev Fitzpatrick made a trip to Spencer and affixed two script decals just under the arm rests of the cab windows. Citing that it was a violation of the historical integrity of the Class J and other N&W locomotives, FireUp611! Chairman and Project Director Preston Claytor resigned in protest. Both men deserve to be respected for their convictions.

Regretfully, Fitzpatrick issued the following statement: "Preston Claytor has resigned and we are moving forward as planned. We are most grateful for all Preston has done and he has done a fine job."

A test run was made to Greensboro and back on Thursday, May 21. Posting an "official...top speed" of 40.3 m.p.h., only minor lubrication problems were encountered, just as they had been on the 611's 1982 post-restoration trip out of Birmingham.

A punch and cake "Send Off Celebration" was held at the North Carolina Museum of Transportation on May 23. Ten lucky donors gave $611 each for 30 minutes of throttle time. Following that up on Thursday, May 28, a "Fired Up Photo Charter" around the museum grounds (including a night session) was conducted by *Trains* Editor Jim Wrinn, a 29-year NCMT volunteer.

At long last the big day – May 30, 2015 – came. Appropriately, the sun rose over the rolling Piedmont at 6:11 a.m. The Greenville dispatcher opened the NCMT switch and a "slow/clear" signal beckoned her onto the main line and 160 miles north, then 53 west. She was on her way home!

Representing the Claytor family on the

The restoration team poses in front of their finished product at Spencer N.C. This was made on 611's 65th birthday, May 29, 2015, the day before leaving for home. *Noah A. Bimrose photo*

train, was Jane Claytor Webster, daughter of Robert B. Claytor. Surprisingly, also aboard was retired NS Chairman David Goode who admitted on-camera in a television interview that it was a mistake to kill the steam program. "I'm glad it's back," he said.

At 1:44 p.m, she felt home rails beneath her feet, as she came around the wye to Woodall Rd. from Montview for a service stop. There, several bus loads of VIPs were waiting to join her for the last lap home.

Stretching his legs in a field was NS Chairman Wick Moorman, who has since retired. "This is great," he said as he surveyed the scene. "It's great for the company and it's a shot in the arm for the employees!"

Taking the throttle out of Lynchburg – just like Bob Claytor so many times before him – Moorman, ran her up, over and down Blue Ridge. "He was grinning from ear to ear!" reported Robert Saxtan.

Especially on N&W territory, her supporters cheered the big, fullback-bodied locomotive along, like an appreciative home crowd supporting their team as it took center-court or ran through a goal post.

Thousands and thousands lined the right-of-way. They were waiting at every country crossing –not just four or five like in the old days, but in throngs of 15 to 20. They found their way back into Montvale, Villamont and Bonsack; at Webster, where they must have shut down production at the General Shale Brick Works – the whole shift was standing on the loading platform, one fan lamented about having his vehicle blocked in for five hours.

At Boaz (pronounced as in the plural of a bow tie), east of Vinton and where the pusher used to couple up, at least 200 people were waving from a field!

Fittingly and at the invitation of Moorman, about two miles out of the old passenger station (Now the Link Museum and Roanoke Valley Welcome Center.), Bobby Saxtan assumed the right-hand seat at Milepost 251. He brought the train all the way by the appreciative, waiting mass of humanity and then – after three short blasts on the whistle – backed her up, making two stops to discharge passengers.

About the honor, Saxtan said: "It was a great thrill. The last time it was sad; this time it was happy. I've never seen this many people.

"She runs as good as she ever did. It's like it never really stopped. I can still close my eyes and find everything!"

Much like the one now more than three decades before, an official "welcome home" ceremony, with Bev Fitzpatrick delivering the keynote speech, took place at the exact same spot where she stopped on August 22, 1982. The trackwork had been altered some, the speed through downtown raised a little and the westward afternoon sun bounced off the flanks of the big 4-8-4 and silhouetted those standing on the new Market Square Pedestrian Walkway above.

Many of the same words were spoken and the emotion of the crowd matched the previous occasion. The FireUp611! Committee and members of the VMT BOD were introduced.

Letting the 611 and her benefactors take the limelight, Wick Moorman stood off to the side and said he was "amazed" at the number of people encountered along the way. "This shows you the power, the attraction of our industry," he observed, before he and David Goode were called to the podium and presented keys to the city by Mayor David Bowers.

Following the ceremony, ground was broken for the new VMT maintenance building at 303 Norfolk Ave. SW and the 611 turned her attention to a NS *"21st Century Steam"* excursion schedule. The 611 was not to rest for long, after a brief pose with recently-arrived Y6a No. 2156 and Class A, No. 1218, she made a deadhead move to Manassas on June 2, then ran two Manassas – Front Royal trips each day on June 6-7. The following weekend, she ran two Lynchburg-Petersburg and return trips on June 13-14. She would deadhead to Spencer with a freight train June 17 for her monthly boiler wash, then deadheaded back to Roanoke on Wednesday, July 1st. To close out her 2015 season, she powered a Roanoke - Lynchburg turn in the morning and a Roanoke - Walton turn in the afternoon, each day for July 3-4-5.

So, converse to a Thomas Wolfe novel, "You *CAN* go home again." And, at least in Virginia, "The more things change, the more they stay the same." Also, if "A cat has nine lives," let's hope No. 611's only on her third, as she gracefully bounds across the Blue Ridge like a big, black and red, gold-trimmed jaguar for generations to come!

••••

The FireUp611 Volunteer Experience

The heart of the VMT FireUp611! movement was vested in the volunteers, who donated over 10,000 man hours during 611's restoration.

At the insistence of the North Carolina Museum of Transportation and Norfolk Southern, all those who wanted to participate had to pay for and pass a background check. If no discrepancies were found, they were refunded their money upon completion of 40 hours of service. Those traveling from out of town or who couldn't commute close enough to sleep in their own beds, also had to foot mileage, meals and lodging expenses.

Once those stipulations were passed, a day-long orientation in Spencer was required. Classes were taught by Preston Claytor and – just like the real railroads – a VMT Safety Rulebook, effective May 1, 2014, was provided.

After completion of the safety orientation, those in attendance were free to go out into the roundhouse and perform work. Some were so physically eager that they could not hide their excitement!

Each brought a special set of skills to the table. Some had been in construction or worked in similar fabrication shop surroundings. Perhaps they were the ones who were the most handy. Since I had been a CSX vice president and had just retired as an Amtrak locomotive engineer, I was given a broom...

In conjunction with doing research and being given free reign to take pictures for the book, I signed up to work the week of Nov. 18, 2014. Among those working with me on some of those days were Larry Stanfill (With over 500 hours, he was faithfully there nearly every Monday and Thursday.), Will Sadler, Jason Huffman, Martin Moorefield, Hunter Phillips, Dick Cudd, Adam McDowell, John Otey, Kevin Gilliam, Richard Moore and Kevin Lusk. Each morning, we reported at 7:30.

Another volunteer was Zac McGinnis. I was surprised to learn that I had worked with his maternal grandfather while a young, Chessie System midnight hostler at Handley, W. Va. Water Service Attendant Gildo Pollestrini was just coming on duty as I was getting off.

Some of the tasks that I participated in while there, were carrying transverse arch tubes to the firebox and back, so that they could be precisely measured for cutting; unloading a heavy crate of arch brick (Even with gloves, I'm still digging sawdust splinters from underneath my fingernails!), lining them by size and making sure the count was right, then carrying them to the firebox and back to make sure that they properly fit over top of the arch tubes; repairing and changing out portable halogen lights, cleaning out the smokebox and a myriad of other mundane chores. I also served as a "go-for" to those working up underneath in the firebox and helped Steve Morgan of AES Manufacture make a device to tighten a bow out of the tacked-in rear flue sheet (after first heating it with a torch), before he expertly welded it into place

Besides the brick arch/transverse tubes, the most exciting aspect that I worked on, in juxtaposition to the overall importance of the project, revolved around flue work. First, we were dispatched to cover those remaining on a flatcar with plastic, in an effort to ward off bird bombing and as a moisture barrier; then the flues were picked up with an overhead crane, moved to a table and the ends ground down (my being left-handed didn't particularly cut it, but Hunter Phillips was especially adept at this task), before wiping them repeatedly with cleaning agents. Next, we built scaffolding as a flue rack, loaded it up with the first of those to be inserted and actually participated in the process of threading the very first one through the front sheet!

On one of the days, in an act of blind, Pocahontas Productions benevolence and as a further contribution to the FireUp611! project, I decided to take everyone (16 in all) to lunch. In retrospect, this was a gesture I came to regret, once the rotund Gary Bensman got up from the table...

The following page is topped with a list of volunteers who gave 40 hours or more. They are sorted by number of hours donated, from highest to lowest.

Volunteers

Cheri George
Charles (Larry) Stanfill
John (Will)Sadler
John Otey
Richard Cudd
Robert Saxtan
Dwight Flinchum
Nathaniel Watts
Duane Leonard
Michael Ridenhour
Richard Moore
Adam McDowell
Hunter Phillips
Zachary McGinnis
Preston Spinks

Kevin Lusk
Jason Huffman
Allan Rearick
Scott Czigans
Edward Mooneyham
Bruce Teerlink
James (Sandy) Alexander
Tim Sposato
Martin Moorefield
Byron (Ron) Dame
Jay Jacobson
John Watso
Jesse Jacobson
Jim Stump
Nick Keller

Brady Sanker
Hugh (Robin) Eanes
Luke Miller
Matthew Janssen
Earl Alsdorf
John Bimrose
Michael Breen
John Barnett
David Podruchney, Jr.
Tyler Sposato
Michael Costill
Trey Davis
Chris Slate
Tony Putnam
Patsy Stump

Supply Side Economics

The following is a list of vendors/suppliers who participated in or, in some significant way, made the FireUp611! project possible. Appropriate comments are made to the right of the directory.

Acheson Foundry	Cast smoke box door dogs.
Acton Rentals	Supplier of on-site office trailer/crew room.
AES Manufacture	Expert boiler and other welding by Steve Morgan; a Chattanooga firm that had participated in TVRM's 4501 restoration.
Age of Steam Roundhouse	Donated labor, sending team of five twice and paid their expenses during critical times of disassembly and reassembly. A gift of Jerry Jacobson.
Airgas	Welding and cutting supplies.
Anderson Tube	Pennsylvania firm that supplied boiler and super-heater tubes.
Applied Metals	Scrapped old, removed flues.
Ashcroft, Inc.	Rebuilt double needle cab gauges.
Axalta Coating Systems LLC	Donated Imron paint (see Part III text).
Backshop Enterprises	Wheatridge Colo. firm of Bernie Watts that rebuilt air compressors.
Coastal Inspection	Ultrasonic testing by Dennis Dunavan.
David Parks	Chrome plating of accessories; of Concord, N.C.
Dayton Phoenix Group	Air brake work.
Diversified Rail Services	Gary Bensman's boiler specialty firm.
Eagle Mechanical LLC	Assisted with boiler and welding work through Hank Marcilio.
Emlex Construction Co.	Vinton, Va.; supplied arch brick and calcium silicate boiler insulation at cost.
Everett Railroad	Donated staybolt cap gaskets; a Pennsylvania steam tourist line.
Fab Designs	Fabricated superheater support brackets.
Fastenal	A source of pipefittings, nuts & bolts and general shop supplies.
Ft. Wayne Railroad Historical Society	Assisted with planning and offered general advice.
Glaze Tool & Engineering	New washout plugs; throttle and cam shaft repair.
Graham White	Cleaning of air brake components; made a major campaign donation (see Part III).

Supply Side Economics, continued

Great Lakes Railcar	Registration of VMT rolling stock (including 611) into AAR (Association of American Railroads) system.
Hendrix Barbeque	"Waist" management.
Historic Machinery Services Inc.	Business of Bob Yuill (see Parts II & III).
Hydraulic Depot	New lube hoses.
Industrial Supply Solutions, Inc.	Local Salisbury, N.C, supplier; mainly pipefittings.
James River Equipment	Rental of four-wheel "gator," to get around and transport materials on site.
Larry Evans	Reproduction of Class J cylinder plates that were presented as favors; of Kenova, W. Va.
McMaster Carr	General industrial supplies.
Metal Locking Services	Buffalo, N.Y. firm that repaired exhaust stand after cracking.
Miller Design	Salem, Va. Volunteer donation of souvenir design work (hats, T-shirts, etc.); produced, drew, laid out and led team in application of stenciling for tender lettering and other parts of locomotive. Pocahontas Productions partner. Co-author of *3 Times A Lady*.
Morton Suggestion	Ordering of promotional items, trinkets, etc.
Nalco	Water treatment; also put author Hensley through college by offering summer vegetation control train employment, further familiarizing him with N&W.
Norfolk Southern Corporation	Made it all possible. Thank you, Wick – again and again and again!
Norfolk & Western Historical Society	Offered/provided original locomotive drawings.
North Carolina Transportation Museum	A most congenial host for 611's third coming, thanks to Director Kelly Alexander and her staff.
Phoenix Specialty Manufacturing	Staybolt gaskets.
Republic Services	Dumpster rental and periodic servicing of same.
Pocahontas Productions	Kenova, W.Va. publishing interest that put together *3 Times A Lady*. Repeated general aggravation of Preston Claytor, Scott Lindsay and Cheri George by co-author Tim Hensley for information.
Riverside Machine	Chattanooga firm that supplied and machined new water glass fittings.
Roberts Upholstery	Reconditioned cab seats.
Rowan Bolt	Supplied what is touted in name, in addition to metal nuts.
Rowan Milk Transport	Supplied 6,000 gallons of boiler water for hydrostatic test in only 20 minutes, instead of wasting valuable time waiting to fill with firehose.
Rowan Precision Machining	Custom machine work by Reggie Hall.
Safety Kleen	Supplied and serviced parts washer.
Steam Operations Corp.	Engineering services by Scott Lindsay's Birmingham, Ala. firm; on-site CMO.
Steam Services of America	Manufactured new staybolts.
Strasburg Railroad	Storied Pennsylvania short line and tourist railroad that supplied new staybolts.
Tennessee Valley Railroad Museum	Cut and applied new cylinder jackets.
Tool Rental Depot	Supplied space heaters for warming of hydrostatic boiler test water to required temperature of between 70 and 120-degrees.
Toomey & Associates	Rebuilt cab gauges.
TTX	New control valves for old tool car (9647).
United Rentals	Rental of, assembly and disassembly of fork lifts with shoot booms that allowed the project to proceed without use of high-rated, overhead cranes.
Vapor Locomotive Company	Inspection of and reassembly of rods and rod bearings by Nat Jansan.
Willets Railcar Services	Body work on tender and locomotive, assistance in painting.

Far left: 611 rides N&W rails for the first time in her third career at Lynchburg, May 30, 2015. *Kenneth L. Miller photo*

Below: The crowds were everywhere, people lining the way and in almost every available viewing spot. Bottom: VMT Executive Director Beverly T. Fitzpatrick on the dais. *Tim Hensley photos*

Above: Three offspring of Pocahontas Division steam passenger engineers out of Bluefield, W.Va. were waiting in Roanoke to welcome the 611 home. From left are Bob Gordon, Sherry Buckland Kelly and Tom Gordon. The men are sons of N.B. "Nebbie" Gordon; the lady is the daughter of L.W. Buckland, who – in true Appalachian fashion – addressed everybody as "Honey." Co-author Hensley made oral history tapes of both railroaders. About his feeling for the Class Js, Tom Gordon said: "Those engines are what fed us. They're what made this country!" *Karen Gordon photo*

Left: Even after all the speeches, the 611 still attracts a crowd waiting to leave the scene. Below: A souvenir ticket for the first run. Bottom: After 56 years, a scene most believed would never again happen: A pose of the "Big Three" in Roanoke, with a Class J under steam! *All: Kenneth L. Miller photos*

The July 3-4-5, 2015 Roanoke-based trips allowed a lot of variety. Top left: 611 has tons of admirers, some of whom have no regard for safety as she rolls west past the original Christiansburg depot. Top right: Shawsvillle, Va. Bottom left: Standing in the murky weather of July 3 at Roanoke, a local shifter passes with a rarity – a caboose!
All: Kenneth L. Miller photos

Fireworks for the Fourth of July! 611 blasts upgrade through Shawsville, Va., proudly carrying American flags while the nearby home proudly displays the Stars and Stripes as well. *Kenneth L. Miller photo*

Running eastbound at Shawsville, Va, the 611, short of the "canteen" and coalboard extensions, looks a lot like it could be 1956, not July 5, 2015. *Kenneth L. Miller photo*

The weather was only moderately cooperative during the weekend, with heavy rain occasionally interfering. It was really pouring down as 611 is headed back to Roanoke at Shawsville. Of interest is the first head car: Norfolk Southern Research and Test Dept. No. 34 is recording dynamometer data on 611 to compare her to tests done 70 years prior! *Kenneth L. Miller photos*

Epilogue

Daddy, Les' go down to the depot...

"Daddy," I used to say on sultry summer evenings when I was but a wee tad nurturing in the sleepy (population 4,320) rail-to-river, coal transportation town of Kenova, W. Va., "Les' go down t' the depot n' watcha' Pow'tan come in!" The former being my father and the latter, of course, was Norfolk and Western's crack *Powhatan Arrow*, which arrived westbound at 6:59 p.m.

"Naw, I don't hardly think so. Not tonight, son," Ralph D. Hensley would say, his thin lips pursing tight with disgust. For he knew what was coming next: An incessant barrage of pleading and whining.

"Aw, go on, take 'im!," Mom would come to my defense from her station at the kitchen sink. While doing the supper dishes, Irene would glare at her hapless husband until he gave in.

So down off winding Barger Hill we would go, soon arriving at the bi-level, yellow-brick Union Station that, in 1957, still served as a regular or conditional stop for 10 passenger trains a day. The downtown of the larger, Chesapeake & Ohio headquarters and hub of Huntington was only eight-miles-distant – the same applied to Ashland, Ky., but the N&W offered a competing alternative to the other big Pocahontas road.

Despite the fact that Grandpa Hensley might chance burbling by at the throttle of a trio of Chessie's new blue and yellow diesels on the lower level, it was the sleek, streamlined 4-8-4s that called upstairs which commanded my attention. Besides, I could still watch for him, looking down on the triple-track C&O from the elevated N&W platform, at an intersecting point where more coal has passed over than any place in the world. Taking the steps two at a time as my nostrils filled with the pungent scent of creosoted wood, I knew the battle was won upon reaching the shelter up top. Surely, dad wouldn't turn back now?

If the old man was in an exceptionally good mood, we might venture out to KX Tower, located across the double-track main at the east end of the platform in the apex of a wye, where the former Twelve Pole and remaining Big Sandy lines split. There, we could listen to Operator Okie Billups bellow "KUH-NO-VIE!" at the tinny voice of the inquiring dispatcher in Portsmouth and watch the little red lights on his CTC board materialize into the Cyclopean headlight of an N&W Class J.

Most of the time, we would wait with the boarding passengers, watching for Train No. 25 to dash around the curve out of the east and come to a squealing stop. The urgent bell, which never stopped ringing, spurred them not to tarry.

Upon several occasions, I was hoisted up into the cab by Engineer Walter Carter. To me, he was the personification of everything that was good and open and honest in my then innocent life and no little influence in the selection of future occupational endeavors. Later, we would become great friends – he lived to be 101; In the 1920s, he boarded at what is now my *Trainmasters House* Bed & Breakfast and I sleep in the same room he did.

Voop! Voop! Two short blasts of her three-chime, stair-stepped whistle announced that the big Class J was ready to go. In an ever increasing, staccato roar, she would charge off across and through the Ohio River Bridge and approach viaduct – longest on the N&W. Suddenly – almost violently, it seemed – the cab would lurch left as she exited the imposing structure and the westward sun would glint off the flanks of her sculpted boiler and tender. Now, only a *loooooww* moan as she warned the crossings of South Point, more than a mile away in the Buckeye State. Then...gone.

So many things change in one's life. A class of engines is scrapped, save one. A man divorces and starts another family. That most sacred of all friends – mother – dies. A boy grows up to be an engineer and a railroad executive, himself.

But the 611 and her class of locomotive have been a constant in mine, even in temporary times of cold steel, something that I could always cling to and depend on to provide comfort in an hour of emotional need. Back when I was a little boy in Kenova, little did I dream that I'd get to run her four times – twice on ferry moves of 150 miles! Thank you, Bob and Bobby.

Memories are, indeed, precious.

Little boys around Kenova didn't think it was the *Powhatan Arrow* unless it had the 606 on the head end. Train 25 is shown stopped at Kenova Union Station on the approach viaduct of the Ohio River Bridge, 1957. Inset: Timmy Hensley (left) and his father Ralph (in shorts). *Bruce R. Meyer photo*